AIR POLLUTION IN INDIA AND ITS IMPACT ON THE HEALTH OF DIFFERENT INCOME GROUPS

AIR POLLUTION IN INDIA AND ITS IMPACT ON THE HEALTH OF DIFFERENT INCOME GROUPS

KAKALI MUKHOPADHYAY

Nova Science Publishers, Inc.
New York

For permission to use material from this book please contact us:
Telephone 631-231-7269; Fax 631-231-8175
Web Site: http://www.novapublishers.com

LIBRARY OF CONGRESS CATALOGING-IN-PUBLICATION DATA

Available Upon Request

ISBN: 978-1-60456-935-3

Published by Nova Science Publishers, Inc. ✢ New York

To

Google,

my beloved nephew

CONTENTS

LIST OF TABLES

LIST OF FIGURES

PREFACE

This book reviews the status of air pollution and its impact on the health of different income groups in one of the developing countries, India. The study estimates the total emissions of CO_2, SO_2 and NOX in India during 1983-84 to 2003-4 using Input-Output Structural Decomposition Analysis. It explores the contribution made by different income groups on the emissions in India. To assess the health impacts of air pollution on different income groups the book also conducts a micro study for a metro city, Calcutta. The micro analysis evaluates the extent of deterioration in air quality, identifies the factors responsible for deterioration, and finally makes an assessment of the impacts of deteriorated air quality on human health of different income groups. One of the biggest tasks at present, the book concludes, is to tackle the generation of emission by the higher income groups along with the improvements of the health scene of the economy. It calls for proper policy for the mitigation of air pollution from the experience of other countries.

FOREWORD

The cumulative environmental impacts of the burning of fossil fuels have created worldwide concerns for their contributions to global climate change. However, in many countries, the local impacts of energy uses are of no less concern. Apart from the industry, households are also major consumer of commercial energy and contribute, and thus, to the total energy use of a country.

Domestic consumption of commercial energy is a major source of emission, the level of which is rising rapidly. In recent years considerable knowledge about the consequences of air pollution on health has been gained, but much of this knowledge has not been made available in a form accessible to stakeholders beyond scientific researchers. The impact of air pollution on health, especially of the poor people is a matter of grave concern to any developing society. It is an important agenda for the academics and policy making community of the developing countries like India. Unfortunately, no systematic and integrated research has been conducted on the diverse avenues through which pollution affects human health in general, and the poor in particular. The present book is an attempt in that direction to make some of this knowledge accessible to the wider stakeholder community.

The book explores the integrated relation of environment and health in general, and role of different income groups in particular. It reviews the status of air pollution and its impacts on health of people belonging to different income groups in India. It estimates the total emissions of CO_2, SO_2 and NO_x in India during 1983-84 to 2003-4 using input-output structural decomposition analysis (SDA). It examines the contribution made by different income groups on the emissions in India. Further, the micro study evaluates the extent of deterioration in air quality, identifies the factors responsible for such deterioration, and finally makes an assessment of the impacts of deteriorated air quality on human health of different income groups.

The results show that the rapidly growing Indian economy is also the source of considerable pollution. Use of cleaner technologies had been introduced, albeit at a slower pace, have not reduced the growth of the overall emission in India. Results of study of the different income groups reveal that higher income groups have mostly adopted such lower emission technologies. This has been possible due to their high level of consumption of energy and ability to invest. However the lower income groups are minor players in this respect.

The micro study on the Indian city of Kolkata reveals how people in the lower income groups suffer more from the pollution led diseases compared to those in the higher income groups who spend more on energy.

The book is based on a study funded by the Indian Council of Social Science Research, New Delhi during 2003-2006. The study was carried out initially at the Centre for Development and Environment Policy when the author was with the Indian Institute of Management in Kolkata. The findings of the study have already been presented in various international conferences. It is hoped that the book will make important contribution to the growth of concern over local impacts of pollution.

Jayanta Bandyopadhyay
President
The Indian Society for Ecological Economics
Delhi, India
19 April 2008

About The Author

Kakali Mukhopadhyay earned her Masters in Economics from Kanpur University, and M.Phil in Economics from Jadavpur University, India. After receiving her PhD degree from Jadavpur University in 2001, she joined the Centre for Development and Environmental Policy, Indian Institute of Management Calcutta, as Post Doctoral Fellow and subsequently as full time Visiting Assistant Professor. She was also a faculty at Madras School of Economics, India. Presently she is a Research Associate at the Department of Agricultural Economics, McGill University, Montreal, Canada. Her research in the past few years has been guided by the integration of energy and environment, trade and environment, air pollution and health, and renewable energy primarily using Input-Output and Multi-Regional Computable General Equilibrium Modeling. She has received a number of international fellowships from World Bank, ADB, Indo-Dutch program, and Ford foundation. She was a visiting fellow at the Faculty of Business Management, Oulu University, Finland; Stockholm Environment Institute, Sweden; SOM Research School-Groningen University, and MERIT-Maastricht University, the Netherlands; and School of Environment Resources and Development (SERD), Asian Institute of Technology, Thailand.

She has published her works in various peer reviewed international and national journals. She has authored two books entitled "Energy Consumption Changes and CO_2 Emissions in India"(2002) published by Allied Publishers, New Delhi, India, and "Trade and Environment in Thailand: An Emerging Economy"(2007) published by Serials Publications, New Delhi, India.

Chapter 1

INTRODUCTION

Environmentalists are very much concerned with GHG concentration world wide. By giving more weightage on it, they have arranged so many negotiations, summits and conferences to discuss about the control of GHG concentration. It is emphasised in the reports on Climate Change 2001 prepared by the Working Group of IPCC to achieve the targets of the Kyoto Protocol (UNFCC) in the short run and stabilize atmospheric concentrations of GHGs in a longer term. It reached an agreement for binding emission targets for GHG emissions reduction with a global average of 5.2% from the industrialised world as a whole by 2012(Europa, 2005). It varies, of course, from country to country.

Human activities are the main source of most of the observed warming of the past half-century and a clear relationship between the growth in manmade greenhouse gas emissions and the observed impacts of climate change is claimed by the policymakers of the IPCC Assessment Report 4(IPCC, 2007). The climate system is more vulnerable to abrupt or irreversible changes than previously thought. Avoiding the most serious impacts of climate change -- including irreversible changes – will require significant reductions in greenhouse gas emissions. Mitigation efforts must also be combined with adaptation measures to minimize the risks of climate change (IPCC, 2007).

Climate change has been induced by green house gases owing to use of fossil fuels in different activities of an economy. The environmental effects of these fuels are of growing concern owing to increasing consumption levels. It is expected to increase considerably in most countries of the Asia pacific region over the next three decades. In addition to global warming acid deposition, ozone depletion, forest destruction is becoming increasingly problematic. Under business-as usual conditions, regional emissions of sulphur dioxide are expected to increase fourfold by 2030 over those of 1990; emissions of nitrogen oxides are expected to increase three fold (GEO, 2000).

1.1. SOURCES OF POLLUTION

Air pollution in urban area comes from a wide variety of sources. The single most important source for the pollutants sulfur dioxide (SO_2), nitrogen oxides (NO_x), carbon monoxide (CO), volatile organic compounds (VOCs), and particulate matter (PM) is generally the combustion of fossil fuels. The burning of fuels for road transport and electricity generation is particularly important. There are three major sources of air pollution in urban areas, namely mobile sources, stationary sources, and open burning sources. An extreme level of indoor and outdoor air pollution is arising from the use of coal and biomass (eg, wood, farm waste, cowdung) for cooking and heating in developing countries. Coal-burning to fuel industry and produce electricity is another major source of air pollution in countries like India and China, where availability of cheap coal is an important factor in industrialisation programme of these two countries. In addition, many of developing countries are facing problem of motor vehicle traffic congestion and associated air pollution. The investigation by Mage et al. (1996) indicates that motor traffic is a major source of air pollution in megacities. Fossil fuels burnt by motor vehicles contribute 90 percent to urban air pollution including lead, carbon monoxide, ozone and suspended particulate matter. In most large urban areas in developing countries exposure to air pollution is becoming severe. Reactions in the atmosphere among air pollutants may produce a number of important secondary air pollutants responsible for photochemical smog and haze in ambient air. The pollutants SO_2 and NO_x are produced by the combustion of fossil fuels, particularly for both stationary and mobile source such as smelters for non ferrous ores, industrial boilers and transportation vehicles. The emission of CO_2 occurs wherever fossil fuels are burnt. Apart from that energy in the form of biomass, containing carbon fixed from the atmosphere, releases CO_2 into the atmosphere.

1.2. Air Pollution and Health

How air pollution and health relates- a snapshot is presented in this section. Later, we describe the health impact considering each pollutant.

The combustion of fossil fuels in industries and vehicles in particular has been a major source of pollution posing health hazards. Concern about the impact of air pollution has led policy community, governments and local authorities across the globe to regulate, among other things, the burning of fossil fuels, industrial effluence, cigarette smoke, and aerosols. This attempt has often considered the findings about the impact of pollution on human health. Attempts are also to detect and quantify pollutants and also to develop urban and rural air pollution networks to monitor levels of atmospheric contamination.

The primary factors that determine the nature of the air pollution problem in cities of developing countries are mixture of dominant emission sources and type and quality of fuel used, while secondary factors are the size of the populated area affected by the emissions, topography and climate of the area (World Bank, 1997).

Source mix and fuel quality determine the amount and composition of emissions. City size, topography, and climate, influence the pollutant concentration level through dispersion (dilution). It also impacts secondary pollutant formation through chemical reactions in the atmosphere. The degree of secondary pollutant formation is also a function of the source mix. These problems extend over continuously growing pollutants hazards and ecosystem degradation (World Bank, 1997).

Air pollution may be defined simply as the presence of substances in air at concentrations, durations and frequencies that adversely affect human health and the environment (McGranahan and Murray, 1999). The remains of early humans demonstrate that they suffered detrimental effects of smoke in their dwellings (Brimblecombe, 1987). Classical writers provide evidence of urban air pollution in the cities of Rome and Athens, and the medieval cities of Europe experienced levels of air pollution considered by the citizens to be unhealthy (Brimblecombe, 1987). Once sulphur and nitrogen compounds have been emitted to the atmosphere, concentrations of gases and acidic deposition impact the local environment. Atmospheric transport and chemical transformation of the pollutants can also form deposition causing impacts far from the point of emission. Nitrogen and Sulphur pollutants cause acidification of lakes and soils and impacts on human health, crop productivity, forest growth, biodiversity and man-made materials.

High levels of outdoor air pollution from stationary and mobile sources contribute considerably to the burden of disease with disorders of the cardio vascular and respiratory systems. In addition, indoor air pollution resulting from domestic use of solid and fossil fuels, particularly in rural areas, accounts for a large number of cases of several diseases(including respiratory infections and chronic obstructive pulmonary diseases) especially among young children and women(WHO, 2002).

So the effect of air pollution on health has become a major concern in recent years. Epidemiological research into air pollution over the past 20 years has demonstrated cardio respiratory health effects ranging from minor respiratory symptoms to increased hospital admissions and mortality. These issues must be addressed seriously if our globe is to achieve a bright energy future with minimal environmental impacts. Evidence suggests that the future will be negatively impacted if the world keeps degrading the environment (Dincer, 2000).

1.3. Features of the Major Pollutants and Likely Impacts on Health

Before going in details of the core problems of air pollution and its impact on health, we brief the features of the air pollutants and the health effects of each of these pollutants.

The main sources of the sulphur dioxide are the combustion of fossil fuels and industrial refining of sulphur containing ores. Depending on the source, the ratio of particulates to SO2 in the ambient air may vary (American Thoracic Society, 1996). In areas where fossil fuels with high sulphur content are used, such as in Beijing, China, high levels of SO2 may be reached, especially during the warm season. Sulphur dioxide is a colourless gas, which can react catalytically or photo chemically with other pollutants or natural components of the atmosphere to produce sulphur trioxide, sulphuric acid and sulphates. SO2 is normally a local pollutant, especially in moist atmosphere, but in oxidized forms it can persists and be transported considerable distances as a fine particulate. It is an important component of acid deposition and haze. Gaseous sulphur dioxide can remain in dry atmospheres for many days and be subject to long range transport processes (Murray and McGranahan, 2003).

Concentrations of more than 10,000 μg/ m^3 SO2 can give rise to severe effects in the form of broncho constriction, chemical bronchitis and chemical tracheitis. Concentrations in the range 2,600-2,700 μg/m^3 give rise to immediate clinical symptoms with bronchospasm in asthmatics. Epidemiological studies indicate the following effects after short term SO2 exposures: possible small reversible declines in children's lung function (250-450μg/ m^3); aggravation of bronchitis (about 500 μg/ m^3); increased mortality (500-1, 000 μg/ m^3) (World Bank, 1997). Sulfuric acid and other sulfates also have human health effects. High concentrations of SO2 (>1,000 μg/ m^3) together with suspended particles are believed to have been responsible for high mortality levels during London smog (The December 1952 smog was associated with 4,000 excess deaths). Several single and multi pollutant time series studies have observed associations between SO2 and daily mortality and morbidity (WHO 2000a; Schwela, 2000).

The major sources of anthropogenic emissions of nitrogen dioxide (NO_2) into the atmosphere are motor vehicles and stationary sources, such as electric utility plants and industrial boilers. Consequently, data on health risks, ambient concentration and standards and guidelines are generally expressed in terms of NO_2 rather than NOx. NO_2 is a reddish brown gas and is a strong oxidant and soluble in water (World Bank, 1997). In most cases, atmospheric nitrogen is oxidized to NO during combustion and then oxidized to NO_2 when emitted into the atmosphere. This process is temperature dependent, with less thermal NO being produced in lower temperature combustion processes. The atmospheric oxidation of NO to NO2 is caused by reaction with O_3 and other oxidants, such as HO_2. This occurs rapidly even when there are relatively low concentrations of NO and oxidants in the atmosphere. Other atmospheric contributions come from non–combustion processes (For example, nitric acid manufacture, welding processes and the use of explosives). Indoor sources of NO2 include tobacco smoking and the use of gas fired appliances and oil stoves.

Annual mean concentrations in urban areas of developing Asia are generally in the range of 2090 μg/ m^3. Levels vary significantly throughout the day. Peaks generally occurred twice daily as a consequence of rush-hour traffic. Maximum daily and 1/2 hour mean can be as high as 400 μg/ m^3 and 850 μg/ m^3, respectively (World Bank, 1997).

NO2 is highly reactive and has been responsible for bronchitis and pneumonia, and also increasing susceptibility to respiratory infections. NO_2 affects both the cellular and humoral immune system, and impairs immune responses. A review of epidemiological studies suggests that children exposed to NO_2 are at increased risk of respiratory illness (Hasselblad et al., 1992). NO_2 has also been associated with daily mortality in children less than five years old (Saldiva et al., 1994). Chronic bronchitis and individuals with emphysema or other chronic respiratory disease may also be sensitive to NO_2 exposures (WHO, 2003). A chart below is showing a possible health impact due to NO_2 exposure.

Health Outcomes Associated with NO_2 Exposure in Epidemiological Studies

Health effects	Mechanism
Increased incidence of respiratory infections	Reduced efficacy of lung defences
Increased severity of respiratory infections	Reduced efficacy of lung defences
Respiratory symptoms	Airways injury
Reduced lung function	Airways and alveolar injury
Worsening clinical status of persons with asthma, chronic obstructive pulmonary disease or other chronic respiratory conditions	Airways injury

Source: Romieu, 1999.

Carbon monoxide (CO) is a colorless, odorless, tasteless gas that is slightly lighter than air. Natural background levels of CO fall in the range 0.01-0.23 μg/ m^3. Levels in urban areas of developing Asia are highly variable, depending upon weather conditions and traffic density. 8–hour mean values are generally less than 20 μg/ m^3 but can be as high as 60 μg/ m^3(World Bank 1997).

The major source of atmospheric CO is the spark ignition combustion engine. Other contributions come from the processes involving the combustion of organic matter (for example in power stations, industry, and waste incineration). In the indoor environment, tobacco smoking can be a significant source of CO (World Bank, 1997).

Carbon monoxide (CO) is one of the most common and widely distributed air pollutants. It is a product of the incomplete combustion of carbon-containing materials, in particular fossil fuel. Carbon monoxide enters the bloodstream and reduces the delivery of oxygen to the body's organs and tissues. People who suffer from cardiovascular disease, particularly those with angina or peripheral vascular disease are much more susceptible to the health effects of CO. The health effects of CO result principally from its ability to displace O_2 on hemoglobin, forming carboxyhemoglobin (COHb).

These health effects are usually related to blood levels of COHb (expressed as a percentage), which can, in turn, be related to exposure (as a function of exposure time as well as concentration). Certain neurobehavioral effects can be expected at about 5% COHb (moderate activity for 8 hours in 40 mg/m^3) that can be related to observable ambient concentrations. These include: impaired learning ability, reduced vigilance, decreased manual dexterity, impaired performance of complex tasks, and chronic respiratory problems, enhanced development of arteriosclerosis and coronary artery disease. The classic symptoms of CO poisoning are headache and dizziness at COHb levels between 10 and 30 per cent. At COHb levels higher than about 30 per cent, the symptoms are severe headaches, cardiovascular symptoms and malaise. High levels (40% or high) may cause strokes, involving unconsciousness and convulsing, brain swelling and protusions, death to part of the brain, comma or even the death of the individual (World Bank,1997; Romieu, 1999).

Various studies in developed countries have documented significant association between daily variations in CO and an increase in premature mortality or hospitalizations from congestive heart failure (Schwartz, 1995; Burnett et al., 1998). Studies revealed that the association between ambient CO and mortality and hospital admissions due to cardio vascular diseases persists even at very low CO levels. This indicates no threshold for the onset of these effects. It is possible that ambient CO may have more serious health consequences than COHb formation, and at lower levels than that caused by elevated COHb levels (WHO, 2002; Schwela, 2000).

The environmental pollutants (SPM, CO, SO_2 and NO_X) have reached levels exceeding allowable levels to be adequate to safeguards health. Much of the world population lives in areas where levels of air pollution exceed WHO guidelines. More than 1200 million people may be exposed to excessive levels of SO_2. More than 1400 million people may be exposed to excessive levels of suspended particulate matter and about 15-20% of population of Europe and North America are exposed to excessive levels of nitrogen dioxide (UNEP, 1991). The situation in developing countries is even worse than that of developed countries. The human health condition in urban developing world is serious especially in the megacities. Here the study captures the status of selected air pollutants in urban megacities of the developing and developed world.

Table 1.1. Annual Means of Ambient Concentration in Several World Mega Cities

City	Population[a] (millions)			TSP ($\mu g/m^3$)	TSP ($\mu g/m^3$)	SO2 ($\mu g/m^3$)	NOx ($\mu g/m^3$)
	1975	2000	2003	1999[b]	1995[c]	1998[b]	1998[b]
Tokyo, Japan	26.6	34.4	35	43	49	18	68
Osaka-Kobe, Japan	9.8	11.2	11.2	39	43	19	63
Mexico City, Mexico	10.7	18.1	18.7	69	279	74	130
New York, USA	15.9	17.8	18.3	23	n/a	26	79
Los Angeles, USA	8.9	11.8	12	38	n/a	9	74
Sao Paulo, Brazil	9.6	17.1	17.9	46	86	43	83
Rio de Janeiro, Brazil	7.6	10.8	11.2	40	139	129	—
Mumbai, India	7.3	16.1	17.4	79	240	33	39
Delhi, India	4.4	12.4	14.1	187	415	24	41
Kolkata, India	7.9	13.1	13.8	153	375	49	34
Shanghai, China	11.4	12.9	12.8	87	246	53	73
Beijing, China	8.5	10.8	10.8	106	377	90	122
Guangzhou, China	4.0				295	57	136
Shengyang, China	4.49				374	99	73
Jakarta, Indonesia	4.8	11	12.3	103	271	30	148
Dhaka, Bangladesh	2.2	10.2	11.6				
Karachi, Pakistan	4	10	11.1				
Metro Manila, Philippines	5	10	10.4	60	200	33	
Kuala Lumpur, Malyasia	1.4				85	24	
Bangkok , Thailand	7.3				223	11	23
WHO Standards				90	90	50	50

Source: World Development Indicators, 2002 and 2003

[a] United Nations Population Division, World Urbanization Prospects, The 2003 Revision. City population is the number of residents of the city as defined by national authorities and reported to the United Nations. Mostly, the city refers to urban agglomerations.

[b] World Development Indicators (2003) published by the World Bank, pp. 168 –169. http:// www.worldbank.org/data/wdi2003/pdfs/table%203-13.pdf. TSP data are for the most recent year available, most are for 1999; SO2 and NOx data are for the most recent year available in 1990 –98. Most are for 1995.

[c] World Development Indicators (2002) published by the World Bank. http://www.worldbank.org /data/wdi2002 /pdfs/table%203-13.pdf.

1.4. Status of Air Pollution in Mega Cities of the World

Health problems associated with urban air pollution have become one of the world's major environmental concerns (UNEP and WHO, 1992; WRI, 1998).

Ambient air pollution in an increasingly urbanized world directly threatens the health of a large fraction of the world's population. There is growing recognition that air-borne emissions from major urban and industrial areas influence both air quality and climate change through out the world. Deteriorating urban air quality affecting the viability of important natural and agricultural ecosystems is becoming particularly acute in developing countries where the rapid growth of megacities (cities having population equal to or more than 10 million) is producing severe atmospheric pollution (Gurjar et al., 2008).

Since there is a paucity of recent data on emissions for the major cities in the world, we portrayed two different tables (tables 1.1 and 1.2) to show the status of emissions. Table 1.1 shows that many major cities in the world have an ambient air pollution concentration exceeding the World Health Organisation (WHO) standard for allowable air pollution levels. It basically shows the historical behaviour of the megacities of the World. Table 1.2 documents some recent trends. It covers the status of air pollution of some major cities in Asia.

Table 1.2. PM_{10} and NOx Concentration in East and South East Asia

	2000	2005	2000	2005
	PM_{10} (micro grams per cubic meter)	PM_{10}(micro grams per cubic meter)	NOx ('000metric tons of CO_2 equivalent)	NOx ('000metric tons of CO_2 equivalent)
China	85	75	556620	566680
Indonesia	120	96	69130	69910
Japan	34	31	26240	23590
Malaysia	27	25	9350	9920
Philippine	48	26	16890	18940
Singapore	44	40	5880	7970
Thailand	79	77	26030	27990
Vietnam	70	61	27110	37470
Bangladesh	162	140	33540	37100
India	93	68	278700	300680
Pakistan	182	120	73630	80040

Source: World Development Indicators (2007) WDI online databases, World Bank .

One interesting feature is observed from table 1.2 that the PM_{10} concentration for all the countries is declining, while NOx is increasing for almost all the countries. NOx emission is high for the countries in South Asia especially in India.

Here the study assesses the sources and status in detail of developing countries covering south and south East Asian countries. We know that mobile and stationary sources are equally responsible for the air pollution, but these differ among countries and within each region of a country.

In Bangkok the mobile sources contributed about 80% of total NOx, 75% of CO and 54% of PM, while stationary source accounts for approximately 96% of the total SO_2 (PCD,2000). Ambient TSP concentrations fluctuated from 1990 till 2004. Roadside TSP has gone down from 500$\mu g/m^3$ in 1990 to 180$\mu g/m^3$ in 2004. Annual NO2 concentration varies between 50 and 60 μg/m^3. Roadside NO2 concentrations are always higher by 20-25 μg/m^3. But SO2 and CO concentrations are showing a declining tendency.

The power plants boilers and industries are the main sources of SO2 in Beijing while mobile sources are the main source of CO (Yu, 2002). Although TSP and PM_{10} concentrations have been decreasing in Beijing in the period 1996-2004, they still exceed their respective class II standards (200μg/m^3 and 100μg/m^3, respectively) (BJEPB, 2005). An annual concentration of NOx is 130 μg/m^3 approximately between 1999 and 2003. The annual SO2 concentrations decreased substantially by more than 50% in the period 1997-2004 and now comply with the class II standards (60μg/m^3)(Schwela et al.,2006).

In Colombo, transport sector was the largest emitter to the overall total emissions for all pollutants for TSP (approx 88%), NOx (82%) and CO (100%) with industry the main contributor of SO2 emissions (93.5%) (Chadrasiri,1999). Ambient PM_{10} value fluctuates around 80μg/m^3 and exceeds the USEPA annual standard (50μg/m^3) as well as the EU standards (40μg/m^3) and the WHO guideline value (20μg/m^3) every year. CO concentration started to increase in 2003. SO2 level is always high because of the power plants are close to Colombo city (Schwela et al.,2006).

A report by the World Bank indicates that concentration of suspended particulate matter (SPM) and airborne lead in Dhaka are higher than ambient air quality standard of Bangladesh and even higher than the WHO guidelines. In particular, the city's average SPM levels are about two times higher than Bangladesh standard of 200μg/m^3 in residential areas and are more than ten times higher than the WHO guidelines of 120μg/m^3 (24 hours) in commercial areas (Brandon, 1997). Another World Bank report (Xie et al., 1998) estimated that Dhaka's motor vehicles emit more than 3,700 tons of particulate matter less than 10 microns in diameter (PM_{10}) per year. Annual averages of PM_{10} are between 110 and 140 μg/m^3 and do not comply with the national standards (50 μg/m^3). $PM_{2.5}$ also have an increasing tendency, it is always higher than the proposed national standards (Schwela et al.,2006). The annual mean concentrations of CO are approximately 1000 μg/m^3. A study by Bangladesh Atomic Energy Commission revealed that the concentration of lead in certain parts of Dhaka is way above the WHO recommended standard (Schwela et al.,2006).

As Vietnam is a rapidly growing economy, the more dominant emission sources in Ho Chi Min city are industry and transport. Industrial plants contribute largely to SO_2 and CO_2 emissions while mobile emissions are the main emitters of CO (84%), NOx (61%) and hydrocarbon (94%) (Vietnam, Registrar, 2002). Motorcycles are a main contributor to emissions of CO, HC and volatile organic compounds while trucks are a major contributor to vehicle related emissions of SO_2 and NO_2. Both trucks and motorcycles are the major emitter of vehicle related CO_2 (Thang, 2004). NO_2, SO_2, and TSP concentrations show an increasing trend in Ho Chi Min City. The mean annual concentrations of SO_2 in Hanoi exceeded the standards, while NO_2 has just met the standards in 2002, after that the concentration shows an increasing trend.

In Hong Kong the public electricity generation sector is a major contributor to emissions of SO_2 (90%), NO_x (76%) and PM (37%). The road transport is a major source of PM (38%) and CO (85%) (EPD, 2004a). A serious problem exists with respect to Pearl River delta affecting air quality in Hong Kong. The PRD emits 95% of PM_{10}, 88% of VOC, 87% of SO_2, and 80% of NO_2 while Hong Kong is responsible for 5%, 12% and 13% and 20% of emissions for these compounds (Civic Exchange, 2004). The concentration of PM_{10} continues to exceed, O_3 also shows a rising tendency, and while NO_2 concentrations have remained relatively constant after 2000 but all are above the EU standards (EPD, 2004b).

In Jakarta mobile sources account for the 80% of the total SO2 emissions and 87% of NO2 emissions while stationary sources (industry, power plant utilities and household) account for 91% of PM emission compared to approx 8% emitted by mobile sources (Wirahadikusumah,2002). More than 50% of NO_x, SO2 and PM_{10} are emitted by the private motor vehicles, buses, and trucks (Wirahadikusumah, 2002). In 2000, the concentrations of most pollutants are above the WHO guidelines, but lead concentrations are decreasing because of the phasing out of lead in vehicle fuel in 2001. Ambient concentrations of O3 have an increasing tendency to exceed the national standards in all years. In 2002 annual concentrations of NO2 exceeded the Indonesia's air quality standard ($60\mu g/m^3$) while CO concentrations remained stagnant (Schwela et al.,2006).

The main sources of air pollutants in the Katmandu valley are motor vehicle exhausts, road re-suspension, brick manufacturing, industries, domestic fuel consumption and refuse burning. While industrial sources of pollution have declined but other sources are increasing with the high rate of urbanization in the valley (KFW, 2004). The SO2 and NO2 concentration are high in 2003 and exceed the WHO standards. But PM_{10} exceeded five times above the WHO and USEPA standard (KFW, 2004).

In 2003 mobile sources in metro Manila were responsible for 85% of PM emissions while stationary sources were responsible for 15% (Anglo, 2004). The food product industry is the largest contributor of industrial PM in Manila city followed by textile products (World Bank 2002). SO2 concentrations have been reduced significantly but $PM_{2.5}$ is not complying with the standards. TSP has also exceeded the guideline value (DENR, 2005).

Fugitive emissions contribute as much as 85% of total TSP emissions and 50% of PM_{10} in Taipei. Emissions from petroleum fuelled vehicles are a source of NO_X (26%), CO (51%), and diesel vehicles are also key source of SO_x (16%) and NO_x (38%). Motor cycles are also a key contributor of CO. NO2 and CO levels are approaching to meet the standards, while SO2 is under control, but O3 is fluctuating from 1994-2004 (DEP, 2003, 2004).

The above assessment shows that most of the developing countries are exceeding guidelines of WHO. Already we have seen that the higher level of pollution is accentuating several pollution led diseases. Thus a significant number of air pollution related health problem is expected to occur in these cities. Long term exposure to air pollution can lead to premature death by increasing the rate at which lung tissue ages, by contributing to chronic obstructive lung disease, and by exacerbating cardiovascular disease. Sudden rise of pollution level (acute exposure), on the other hand, can cause the people with history of cardiopulmonary diseases or simply weak or susceptible to die prematurely. This is known as the short-term or acute effect.

In a comparative risk assessment of global human health risks, the WHO ranked urban outdoor air pollution as a tenth leading cause of premature deaths and indoor air pollution as a fourth leading cause(WHO,2002). The WHO estimates that 537000 persons die prematurely each year in the South East Asia and western pacific regions due to exposure to urban air pollution. Approximately 1.06 million die annually from exposure to indoor air pollution in these regions. In additions, millions of people are affected by respiratory diseases, which are either caused or aggravated by exposure to ambient air pollution (WHO, 2002). The WHO estimates that 2/3rd of the world's statistics in death and lost life years due to urban outdoor air pollution occur in Asia (WHO, 2002). These estimates are however, based largely on exposure response relationships derived in Europe and North America before 2000 and extrapolated for Asia(WHO, 2000a; 2000b).

The evidence from developed nations can substantiate the harmful effects of air pollutants. Studies in the America and Europe in recent years on the health effects of ubiquitous air pollutants, such as particulate matter and ozone, have documented responses proportionate to exposures, including excess daily and annual mortality, hospital admissions, lost time from school and work, and reduced lung function. These effects constitute a significant public health challenge in developed countries (Lippmann, 2003).

Reports from the developing countries show a causal relationship between air pollution and health effects. Various studies documented increased mortality and visits for respiratory emergencies associated with particulate pollution (particularly with particulate smaller than 10 mm and 2.5mm). Reports also show higher frequencies of respiratory symptoms and low pulmonary function in subjects exposed to particulate. Asthmatic populations appear to be more susceptible to the impact of particulate and SO_2 exposure. The health effects of O_3 have focused on short term exposure and have documented increase in emergency visits and hospital admissions due to respiratory diseases, increase in respiratory symptoms and temporary lung function decrements. Time series studies evaluating associations between O_3 and daily mortality based on limited data suggest that CO exposure is prevalent and may be associated with intrauterine death. Most evidence suggests that populations living in cities with high levels of air pollution in developing countries experience similar or greater adverse effects of air pollution (Romieu, and Hernandez,1999). The next section focuses on the relationship between pollution, health and poverty.

1.5. Pollution, Health and Poverty

The interrelationship between pollution, health and poverty in Asia is important especially in areas with high air pollution and large population at the lowest socio economic level. While there have been many studies which document the overall health impact of air pollution in Asian cities, there have been a very few studies on the relationship between air pollution, health and poverty in developing countries in general, and in Asia in particular (HEI, 2004).

A study by Smith et al. 1999 demonstrates that around 40 to 60 % of acute respiratory infections are due to environmental causes. They added that the adverse effects of air pollution depend on the level of exposure, the population structure, the nutritional status and the lifestyle. They observed that the effects are higher in developing nations than in developed ones, because the number of people living below the poverty line is higher in developing countries. Poor people are often seen as compelled to exploit their surroundings for short-term survival, and are most exposed to natural resources degradation.

There is growing evidence that the poor are affected relatively more by ambient air pollution due to greater exposure, weaker biological defence mechanisms against air pollution, inadequate nutrition and limited access to medical care (Martin et al., 2004;Gouveia and Fletcher,2000; Stern, 2003).

At the same time, air pollution could aggravate the conditions of poverty. It is also expected that the poor will not seek medical assistance quickly, nor be able to afford medicines and treatment. In Hong Kong it is estimated that the poor suffer increased exposure to air pollution and those in low income areas are five times as likely to be hospitalised as those residing in high income areas (Stern, 2003). The economic consequences of an increased burden of disease due to air pollution including cost of illness and loss of income could be substantial.

Cities in developing nations are increasing rapidly in size and diversity. Increasing emissions from vehicular traffic, industry, domestic heating (in temperate climates), cooking, and refuse burning all pose potential risks for large air pollution exposures. The rapidity of economic development combined with the lack of adequate emission controls, makes Asia's megacities prone to more serious air pollution problems than similar cities in industrialized nations. Although the current fossil fuel use in developing countries is still half that of developed countries, it is expected to increase by 120% by the year 2010. If control measures are not implemented, it has been estimated that by the year 2020 more than 6.34 million deaths will occur in developing countries due to ambient concentrations of particulate air pollution (Romieu, and Hernandez, 1999).

Indian scenario is alarming. The next section deals with that.

1.6. Indian Scenario

Deterioration of India's air quality has been accentuated by industrialization and urbanization. While India's gross domestic product has increased 2.5 times over the past two decades, vehicular pollution has increased eight times, while pollution from industries has quadrupled. Household energy consumption also cannot be ignored in this respect. Households are a major consumer of energy and contribute, to a large extent, to the total energy use of the nation. At present, the share of direct energy use of households in India is about 40% of the total direct commercial and non-commercial indigenous energy use (Pachauri and Spreng, 2002). If, in addition, one takes into account the indirect or embodied energy in all goods and services purchased by households, then about 70% of the total energy use of the economy can be related to the household sector, the remaining 30% comprise the energy requirements of government consumption, investments and net imports (Pachauri and Spreng, 2002). The distribution of population with regard to energy consumption also shows that over 60% have a per capita total household energy requirement of less than 0.5kw per year. In addition to the wide disparities in the quantities of energy used, there are large variations in the types of energy used and pattern of consumption among households.

During the past few decades, India has experienced major changes in its energy consumption patterns - both in quantitative and qualitative terms. This is due to population growth and increase of economic activity and development. As mentioned household sector is one of the major users of energy in India. The pattern of household energy consumption represents the status of welfare as well as the stage of economic development. As the economy develops, more cleaner energy is consumed. Moreover, household energy consumption pattern is likely to vary with the income distribution and its change overtime. Household energy consumption is expected to increase in future along with the growth in economy and rise in per capita income. The projected increases in household energy consumption are expected to result from changes in lifestyles (Pachuri, 2004).

Pollution may be indoor or outdoor and is closely linked with the health. So a study is needed to quantify the air pollution generated from fossil fuel combustion and its contribution by different income groups in India. That will provide some idea of which income groups in India are generating more pollution and who suffer actually from pollution related health diseases.

A survey by Central Pollution Control Board India (CPCB, 2000) has identified 23 Indian cities to be critically polluted. 12 major metropolitan cities in India produce 352 tonnes of oxides of nitrogen, 1916 tonnes of carbon mono oxides from vehicular emission and 672 tonnes of hydrocarbon. The CO, SO_2 and NO_x in the ambient air of India are above the WHO safe limit. WHO annual mean guidelines for air quality standards are 90 micrograms per cubic meter for total suspended particulates, and 50 for sulphur dioxide and nitrogen dioxide (World Development Indicators, 2000). The total urban air pollution of SO_2 and NOx from major cities in India are 300 micrograms per cubic meter and 250 microgram per cubic meter during 2004 (World Development Report, 2005). Deterioration of air quality is a problem. A majority of the 300 million urban Indians, about 30% of India's population is directly experiencing this (Gurjar et al., 2008). Recently an Environment and Forest Ministry Report, Government of India, released on September 14,2007 has identified 51 cities that do not meet the prescribed Respirable Particulate Matter(RSPM) levels, specified under the National Ambient Air Quality Standards(NAAQS). In 2005, an Environmental Sustaibility Index (ESI) placed India at 101st position among 146 countries(Environmental Pollution in India,2008).

According to UNEP (2001) report, in most of the Indian cities the mean of average values of SPM for nine years (1990–98) ranged between 99 and 390 $\mu g/m^3$ in residential areas and between 123 and 457 $\mu g/m^3$ in industrial areas, exceeding the annual average limit of SPM for residential areas 140 $\mu g/m^3$ and for industrial areas 360 $\mu g/m^3$.

According to the World Health Organization, the capital city of New Delhi is one of the top ten most polluted cities in the world. Apart from Delhi, the other major polluting metros are Mumbai and Calcutta. Nine ambient air quality monitors operate in Delhi (CPCB, 2001) including five industrial and four residential sites (TERI, 2001). Most of the monitoring stations measured TSP, SO_2, NO_2, PM_{10}, lead, benzo-pyrene, and O_3 regularly at a major traffic intersection (CPCB, 2001, 2004). Annual averages of SO_2 and NO_2 often exceeded national standards of 15 $\mu g/m^3$ from 1994 to 2003. NO2 concentrations are increasing since 2001. The mean 24-hr NO_2 levels exceeded the national standard of 30 $\mu g/m^3$ at 8 locations out of 18 (TERI, 2001a). Further, annual and monthly averaged TSP levels (CPCB, 2001) have exceeded the national standards always. While ambient SO_2 levels show a decreasing trend in Delhi (as expected after the introduction of low-sulfur fuel). Table 1.5 shows that ambient CO concentrations in Delhi have consistently violated the CO standard of 2000 $\mu g/m^3$ for residential areas (TERI, 2001). Varshney and Aggarwal (1992) and Singh et al. (1997) observed that 1-hr average O_3 concentrations are exceeding the prescribed WHO standard of 100 $\mu g/m^3$ at various locations in Delhi compared to other large Indian cities such as Mumbai, Chennai, and Calcutta.

Table 1.3 reports the concentration of SO_2 and NO_2 in major cities in India. The RSPM and NO_2 concentration is high for Delhi and Mumbai.

The accumulation of air pollutants in Delhi during winter is more critical (Delhi Pollution Control Committee: New Delhi, 2003). Several emission inventories have been developed for Delhi (Gurjar et al., 2004 and Molina and Molina, 2004).

Table 1.4 documents vehicular emissions in Delhi and their increases relative to base year 1990–1991. Within the past decade, emissions were doubled for SO_2, and increased by 6-fold for NOx, CO, and HC, and nearly 12-fold for TSP. The CO concentration is really high in Delhi (table 1.5) and vehicular emission is partly responsible for it (table 1.4).

Table 1.3. Air Quality in Major Metro Cities (hourly status)

City (Location)	Concentration (μg/m^3)		
	Sulfur dioxide (SO_2)	Nitrogen dioxide (NO_2)	Respirable Suspended Particulate Matter (RSPM)
National Standard	80	80	100
Chennai	6	18	92
Delhi	7	83	303
Mumbai	48	133	301

Source: Central Pollution Control Board, New Delhi, 2007.

Table 1.4. Vehicular Emissions (tons per day) in Delhi

Pollutant	SO_2	TSP	NOx	CO	HC
1990–91	6–10	1–19	44–139	243–492	82–200
1995–96	14–15	26–28	120–397	373–781	123–493
2000–01	18	35–196	261–860	447–4005	156–1542
Average decadal increase factor	2.2	11.6	6.1	6.1	6

Source: Gurjar et al., 2004.

Table 1.5. Ambient CO Trends (1995–2000) in Delhi

Location of Measurements	Annual Average CO Concentrations (μg/m^3)					
	1995	1996	1997	1998	1999	2000
Residential area average (Siri Fort)	No data	No data	3177	3340	3578	2376
Traffic junction average (ITO)	3916	5587	4810	5450	4241	4686

Source: Central Pollution Control Board, New Delhi 2007

Similar to Delhi, the industrial sources are responsible for approximately 48% of the air pollution in Calcutta with mobile sources accounting for 50% and domestic sources 2% (WBDOE, 2003). Large and medium industries emit approx 56% of PM whereas the small units emit 40% of PM consuming only 6% of total coal used in the industrial sector in Calcutta everyday(Chakraborti,2003). Major sources of $PM_{2.5}$ are diesel engine and gasoline emissions, road dust, resuspension, coal and biomass burning (World Bank, 2004).

Table 1.6. Daily Ambient Air Quality in Calcutta, 24 hours (2006)

Station Name	SPM	RPM	SO_2	NO_X
	μg/m³	μg/m³	μg/m³	μg/m³
Calcutta Average	361	219	24	117
Residential Standard	200	100	80	80

Source: West Bengal Pollution Control Board, Calcutta, 2007.

Since 1997 till 2004, there has been a decreasing tendency of TSP and RSPM in Calcutta. Table 1.6 shows the daily ambient air quality of Calcutta. NO_2 concentrations have exceeded the annual standards for residential areas in 2001 and 2002. Significant reduction has been observed for SO_2 concentrations. Lead concentrations are under control (WBPCB, 2003).

In Mumbai the major sources of TSP were identified as being resuspension of road dust, wood combustion, domestic refuse burning and diesel vehicle exhaust. The main source of SO_2 emissions was industrial fuel oil combustion and power plants (82%) and industrial fuel combustion were also major source of NOx emissions (52%). Stone crusher has been identified as a major source of TSP (NEERI, 2004). Annual levels of TSP fluctuates around 250 $\mu g/m^3$ and do not comply with the Indian standard 140 $\mu g/m^3$. PM_{10} levels are also showing a continuous increase. The annual average of NO2 and SO2 declined from 1998-2004.

It is needless to say that pollution of urban air is likely to have a serious impact on the health of the community. The patterns of disease and death exhibited in India strongly suggest the possible importance of environmental factors in today's Indian health scene. Particulate pollution on its own or in combination with SO_2 leads to an enormous burden of ill health, causing at least 500,000 premature deaths and 4-5 million new cases of chronic bronchitis each year in India (World Bank Report,1992). Recently Centre for Science and Environment report (CSE,2006a) said that 60 million people constantly suffer from indoor and outdoor air pollution related diseases and 2.3 million from the burden of peri-natal ailments.

Incidence of poverty is high in India and about one third of the population is below the poverty line. This section of population is likely to be affected by environmental hazards. In conditions of poverty, where more environmental hazards are present in the nearby living and working environment, people are more vulnerable to exposure and effects of harmful pollutants. The environmental threats facing poor people tend to be more directly hazardous to human health. We know that pollution related health hazards are not uniform in all income groups. It affects the lower income groups more than the upper income groups (Lioy, 1990). A million of people die from air pollution in the urban environment, and it is likely that here too the poor suffer most. India is second only to sub-Saharan Africa in terms of deaths and lifetime shortened by disease (CSE, 2006a).

Estimates of the full loss of healthy life due to different causes are reported in World Development Report in terms of Disability Adjusted Life Years (DALYs) lost. According to these estimates India accounted for 392 million DALYs lost in the year 2000, or slightly over 21% of the global burden of disease. Diseases that are typically associated with environmental pollution loom large as causes of India's DALYs losses. Rough estimates indicate that these diseases are responsible for almost 30% of India's total DALYs losses (details in chapter 2).

Recently, the World Health Organization (WHO) is asking governments around the world to improve air quality in their cities in order to protect people's health. WHO revised its existing Air quality guidelines[1] (AQGs) for Europe and expanded these to produce the first guidelines which are applicable worldwide[2]. These global guidelines are based on the latest scientific evidence and set uniform targets for air quality for all regions of the world which would protect the large majority of individuals from the effects of air pollution on health. These guidelines propose progressive interim targets and provide milestones in achieving better air quality. WHO opines that reducing levels of one particular type of pollutant (known as PM_{10})[3] could reduce deaths in polluted cities by as much as 15% every year. The Guidelines are substantially lower than the recommended limits of ozone and sulphur dioxide. These targets are far tougher than the national standards currently applied in many parts of the world. As a result some cities of the world would likely to reduce current pollution levels by more than three-fold.

Air pollution is estimated to cause approximately 2 million premature deaths worldwide per year. More than half of this burden is borne by people in developing countries. In many cities, the average annual levels of PM_{10} (the main source of which is the burning of fossil and other types of fuels) exceed 70 micrograms per cubic metre. The new Guidelines say that, to prevent ill health, those levels should be lower than 20 micrograms per cubic metre (WHO, 2006).

Air pollution, in the form of particulate matter or sulfur dioxide, ozone or nitrogen dioxide, has a serious impact on health. For example, in the European Union, the smallest particulate matter alone ($PM_{2.5}$) causes an estimated loss of statistical life expectancy of 8.6 months for the average European. While particulate matter is considered to be the main air pollution risk factor for human health, the new Guidelines also recommend a lower daily limit for ozone, reduced from 120 down to 100 micrograms per cubic metre. Achievement of such levels will be a challenge for many cities, especially in developing countries, and particularly

---

[1] "Building upon the work carried out for several years on air pollution, WHO has now set new targets which Member States can refer to in setting policy. The countries can measure their distance to these objectives, estimate the health impact of current pollution levels and benefit from health gains by reducing them" (WHO Regional Office for Europe, coordinating the process of the Guidelines' update from the WHO office in Bonn, 2006). These new guidelines have been established after a worldwide consultation with more than 80 leading scientists and are based on review of thousands of recent studies from all regions of the world. As such, they present the most widely agreed and up-to-date assessment of health effects of air pollution, recommending targets for air quality at which the health risks are significantly reduced.

[2] Many countries around the world do not have regulations on air pollution, which makes the control of this important risk factor for health virtually impossible. The national standards which do exist vary substantially, and do not ensure sufficient protection for human health. While the World Health Organization accepts the need for governments to set national standards according to their own particular circumstances, these guidelines indicate levels of pollution at which the risk to health is minimal. As such, the new WHO guidelines provide the basis for all countries to build their own air quality standards and policies supporting health with solid, scientific evidence.

[3] "By reducing particulate matter pollution from 70 to 20 micrograms per cubic metre as set out in the new guidelines, we estimate that we can cut deaths by around 15%," said Dr Maria Neira, WHO Director of Public Health and the Environment. "By reducing air pollution levels, we can help countries to reduce the global burden of disease from respiratory infections, heart disease, and lung cancer which they otherwise would be facing. Moreover, action to reduce the direct impact of air pollution will also cut emissions of gases which contribute to climate change and provide other health benefits." (WHO, 2006).

those with numerous sunny days when ozone concentrations reach the highest levels, causing respiratory problems and asthma attacks(WHO, 2006).

For sulfur dioxide, the guideline level was reduced from 125 to 20 micrograms per cubic metre. Experience has demonstrated that relatively simple actions can rapidly lower sulfur dioxide levels and directly result in lower rates of childhood death and disease. The guideline level for nitrogen dioxide remains unchanged. Meeting these limits, which are essential to prevent the health consequences of exposure remains a great challenge in many areas. The WHO guideline values are given in table 1.7.

There are several attempts already been made to estimate the air pollution and health impact in different parts of the world. The impacts of air pollution on health have been studied in North America and Europe for many decades, but research on effects on health in developing countries is less advanced.

Table 1.7. Guideline Values by WHO in October, 2006

$PM_{2.5}$	PM_{10}	O_3	NO_2	SO_2
10µg/m^{3}annual mean	20µg/m^{3}annual mean	100µg/m^3 8 hour mean	40µg/m^3 annual mean	20µg/m^3 24 hour mean
25µg/m^3 24 hour mean	50 µg/m^3 24 hour mean		200µg/m^3 1hour mean	500µg/m^3 10-minute mean

Source : WHO, 2006.

As the impacts on health associated with energy use are inevitable as development is linked with the energy consumption levels leading to high emissions. Realizing the need to control and regulate emission of pollutants, the present study concentrates on the above issues in India.

We have some scattered studies in India especially on metro cities. But none of them covers the assessment of air pollution, generation of pollution by different income groups as pollution and poverty are closely linked, and finally who are the ultimate sufferers from pollution. Moreover, a comprehensive epidemiological assessment of the situation has not been done in India so far. The present study addresses in a modest way the status of the air pollution and its generation by different income groups in India and assesses the impact on health of different income groups in Calcutta, a metro city of India. The *objectives* of the present study are:

a) To estimate the emissions of CO_2, SO_2 and NOX in India during 1983-84 to 2003-4;
b) To investigate the changes in emissions and various sources of changes in CO_2 SO_2 and NO_x emissions using input-output structural decomposition analysis (SDA) during the period 1983-84 to 2003-4;
c) To find out the contribution made by the different income groups on emissions for the period 1983-84 to 2003-4;
d) To carry out a case study for Calcutta, a megacity, to assess the impacts of deteriorated air quality on human health in order to corroborate the macro findings.

Approach of the Book

A basic introductory note is portrayed in Chapter 1, which deals with the severe problems of air pollution and how it affects health. Chapter 2 reviews the literature on the status of air pollution and its impact on health in developing countries, especially India. The formulation of the Model is furnished in Chapter 3. It develops the methodology to estimate the overall emissions in India and to identify the responsible factors and the contribution of the different income groups. The detail description of different sources of data used in application of the model and processing of data is featured in chapter 4. Chapter 5 discusses detailed empirical findings. The total emission changes and factors responsible for these changes in India during 1983-84 to 2003-4 are analysed along with the generation of pollution by different income groups in the country. Chapter 6 deals with the micro study based on Calcutta. It covers detailed description of the study site, and survey methodology. It also provides features of the surveyed household in light of the different income groups in each area. Chapter 7 investigates the health impact of air pollution using simple logit model for deriving the relationship between pollution, energy expenditure and general health conditions. It also includes the literature on the economic valuation on health impact of air pollution. Chapter 8 concludes the book with policy implications.

Chapter 2

STATUS OF AIR POLLUTION AND ITS IMPACT ON HEALTH IN INDIA AND OTHER DEVELOPING COUNTRIES

In many developing countries economic growth without adequate environmental protection has resulted in widespread environmental damage, creating new environmental problems. Populations in urban areas are at risk of suffering adverse health effects due to rising problems of severe air and water pollution.

Air pollutants are likely to have adverse effects on different human populations, depending on the range of exposure, different pollutants, the population structure, the nutritional status and the lifestyle. It is observed that the potential health effects of air pollution may be even greater in developing nations than those reported for developed nations. In this chapter, the epidemiological studies describing health effects of air pollution in the developing countries are reviewed briefly. This review basically covered the studies that have evaluated health effects in relation to exposure to the critical air pollutants[4] like particulate matter, sulfur dioxide, nitrogen dioxide, ozone, carbon monoxide and lead.

The review divides the literature on air pollution and health into three major sections. The first section contains the health effects of major air pollutants along with some scattered epidemiological evidences in different countries. It emphasizes the causal link between air pollution and its impact on health for developing countries primarily. The second section deals with the impact of pollution on health relating to India only. The literature concentrating on the valuation of health cost from air pollution in India and other developing countries is presented in the third section.

2.1. THE EVIDENCE OF AIR POLLUTION IMPACTS ON HEALTH

There are several studies in developing countries that have established the linkage between environmental pollution and its impact on health. Before describing the extensive literature on developing countries, we present one or two examples from world wide

[4] The review of literature is not confined to CO_2, SO_2 and NOx, as the current study estimates. It covers the pollutants as much as possible across different countries in the world.

experiences. Most of the studies are dealing with particulate matter (PM)[5] and ozone (O_3), a serious concern in the developed world.

Two cohort studies (Dockery et al., 1993; Pope et al., 1995) conducted in the US have reported a large mortality estimate related to long term exposure to fine particulates. These estimates suggested an increase in mortality from 17 per cent to 26 per cent over a range of approximately 20μg/m^3 of $PM_{2.5}$. In addition, post neonatal mortality has been associated with exposure to PM_{10} during the first two months of life. An increase of 25 per cent in overall post-neonatal mortality was observed for 30μg/m^3 range of PM_{10} concentrations (Woodruff et al., 1997). Samet et al., (2000) summarizing data from 20 cities in the US, reported an increase in total mortality of 0.51 per cent per 10μg/m^3 of PM_{10}. For cardiovascular mortality, this estimate reached 0.68 per cent. Daily mortality appears to be more strongly associated with concentrations of $PM_{2.5}$ than with concentrations of larger particles[6] (Schwartz et al., 1996; Klemm et al., 2000). A longitudinal data from the Children's Health Study, conducted in 12 communities of Southern California, suggest a significant deficit in lung growth related to NO2 and fine particulate exposure (Gaudermann, 2000).

Lippmann (2003) analyses several studies in USA and Europe in recent years on the health effects of ubiquitous air pollutants, such as particulate matter (PM) and ozone (O_3). These studies documented responses proportionate to exposures, including excess daily and annual mortality, hospital admissions, lost time from school and work and reduced lung function. Ozone levels in many developing countries are currently lower than in the USA and Europe. Precautionary measures on sources of hydrocarbons and nitrogen dioxide can be instituted to keep them under control to minimize adverse effects.

Man-made outdoor air pollution in the world's cities, derived largely from combustion processes, is a complex mixture with many toxic components. Cohen et al. (2005) have indexed this mixture in terms of particulate matter (PM), with serious health effects. They estimate that ambient air pollution, in terms of fine particulate air pollution ($PM_{2.5}$), causes about 3% of mortality from cardiopulmonary disease, about 5% of mortality from cancer of the trachea, bronchus, and lung, and about 1% of mortality from acute respiratory infections in children under 5 year, worldwide. This amounts to about 0.8 million (1.2%) premature deaths and 6.4 million (0.5%) years of life lost (YLL). This burden occurs predominantly in developing countries. In Asia the figure is as high as 65%. These estimates consider only the impact of air pollution on mortality (i.e., years of life lost). If air pollution multiplies both incidence and mortality to the same extent (i.e., the same relative risk), then the DALYs for cardiopulmonary disease increase globally by 20%.

The health impact of PM_{10} and ozone in 13 Italian cities (WHO, 2006a) shows that average PM_{10} levels in Italian cities in 2002-2004 ranged from 26.3 to 61.1 μg/m^3. The health impact of air pollution is large: 8220 deaths a year, on average, are attributable to PM_{10} concentrations above 20 μg/m^3. This corresponds to 9% of the mortality for all causes (excluding accidents) in the population over 30 years of age. The impact on short-term mortality is 1372 deaths. It constitutes 1.5% of the total mortality in the whole population.

[5] Exposure to PM has been associated with a wide range of health effects, but its effects on mortality are the most important, and should be considered for global assessment.Levels of particulate matters in the air in the cities of the developing countries, especially those due to coal smoke, can be high and the adverse health effects they produce can decrease as emissions are reduced.

[6] This has special implications for the developing countries where vehicular traffic with poorly maintained engines and extensive use of diesel fuel is a major source of particulate pollution.

Hospital admissions attributable to PM_{10} are of a similar magnitude. The impact of ozone at concentrations higher than 70 $\mu g/m^3$ amounts to 0.6% of all causes of mortality. Higher figures were obtained for the effects on health that result in morbidity. This study confirms the findings from several investigations worldwide and suggests that reducing emissions from motor vehicles (the main source of PM_{10} in cities of industrialized countries) would benefit the health of urban populations.

The pollution related diseases are common in both world, but to tackle the problem is easier in the developed world than developing one burdened by the population size and large number of poor people.

Here are few epidemiological evidences across major air pollutants among the developing countries. Romieu and Avila (2003) have reviewed the epidemiological studies that describe the health effects of various air pollutants (including particulate matter, sulfur dioxide, nitrogen dioxide, ozone, carbon monoxide and lead) for different cities of the developing world. Reports from the developing nations show similar effects observed throughout the world, adding support for a casual relationship between air pollution and health effects. They have reviewed various studies that have documented increased mortality, visits for respiratory emergencies, higher frequencies of respiratory symptoms and low pulmonary function associated with particulate pollution (particularly for SPM smaller than 2.5 microns). Young children appear to be at higher risk for acute respiratory infections leading to increased morbidity and mortality given the presence of generally poor environmental condition and nutritional status in developing countries. Asthmatic populations appear to be more susceptible to the impact of particulate, sulfur dioxide and ozone exposure. They have concluded that the chronic exposure is of major concern in developing countries given the high levels of air pollutants.

Studies associating daily variation in air pollution level with mortality rates, crude or disease specific, have become quite popular. Such studies, however, can only measure the effects of acute exposure to air pollution on premature deaths and/or morbidity. As these studies cannot capture chronic effects of exposure to air pollution, their estimates should be considered as lower bound estimates of reduction of premature deaths due to drop in air pollution level. A large number of studies came out in last few years using the methodology, mostly from developed countries[7]. Almost all of them found strong positive correlation between level of air pollution and mortality/morbidity rates. These studies have reported their estimates of Dose-Response Functions (DRF)[8]. The easiest and widely followed method to estimate health impact has been to use existing dose response functions. However, these functions have been estimated in developed country settings, and the results may not be easily transferable to the developing countries. Ostro et al. (1995) developed DRFs using developing country data. They estimated the relationships between PM_{10} and total mortality, female mortality, cardiac-specific and respiratory-specific mortality of Santiago, Chile. Running simple OLS regression, they found that strong positive relationship exists between total

---

[7] Fairley(1990), Pope et al.,(1991), Pope et al., (1992), Schwartz(1991), Schwartz and Dockery (1992a), Schwartz and Dockery (1992b), Schwartz (1993). The meta studies of air pollution by Pearce (1996) resulted in the use of transferable dose-response coefficients whereby the statistical relationship between air pollution and human health is applied. It suggests that some forms of air pollution, notably inhalable particulate matter and ambient lead, are serious matters for concern in the developing world.

[8] DRF means the percentage of mortality or morbidity rate in a community due to unit drop (usually $\mu g/m^3$) in air pollutant such as TSP or PM_{10}.

mortality and PM_{10} concentrations. Using Poisson regressions, they found similar relationship between PM_{10} concentrations and other measures of mortality such as cardiac-specific and respiratory-specific mortality. Colder days apparently have strong positive relationship with the number of deaths in all of those regressions.

Mortality studies conducted in developing countries are consistent with estimates from the US and Europe (Borja-Aburto et al., 1997; Hong et al., 1999; Tellez Rojo et al., 2000). A major concern has been raised recently regarding the increased infant mortality linked to particulate exposure in Brazil, Mexico and Thailand (Loomis et al., 1999; Ostro et al., 1999; Gouveia et al., 2000; Conceiao et al., 2001). A summary estimate of these time-series studies suggests that an increase of 10μg/m^3 of PM_{10} could be associated with an increase close to 1 per cent in total mortality for respiratory causes in children less than five years of age. Various studies conducted in Brazil, Chile, Cuba and Mexico have shown an increased risk of respiratory infections in young children exposed to particulate pollution (Tellez-Rojo et al., 1997; Pino et al., 1998; Ilabaca et al., 1999; Gouveia et al., 2000; Romero et al., 2004). The fact that exposed children in developing countries often suffer from additional risk factors such as poor living conditions and nutrition deficiency which increase their susceptibility to the adverse effects of particulate pollution.

Significant exposure-response relationships between maternal exposure to SO_2 and to total suspended particulates (TSP) and low birth weight were observed in studies conducted in China (Wang et al., 1997) and the Czech Republic (Boback et al., 2000). Relationships are also found to exist between these pollutants and fetal growth retardation (Delmeek et al., 1999) and pre-term birth (Ritz et al., 2000).

Studies conducted in Mexico City, where ozone levels frequently exceed by a large margin the WHO guidelines, have documented an increase in asthma-related emergency visits, and an increase in respiratory symptoms in asthmatic children. In addition, studies also found an association between ozone exposure and school absenteeism for respiratory illnesses, even at levels of exposure that are common in many urban areas (average 20 to 50 ppb, 10 am to 6 pm) (Romieu et al., 1997).

NO_2 has also been associated with daily mortality (Saldiva et al., 1994) and intrauterine mortality levels in children less than five years old in Sao Paulo, Brazil (Pereira, 1998). In these reports, NO_2 was more significantly associated than the other pollutants that were studied. Studies reported that CO exposure may be associated with intra-uterine death (Pereira et al., 1998) and with pre-term birth (Ritz et al., 2000).

A World Bank study showed an association between respiratory disease and PM_{10} concentrations. The effect of PM_{10} on daily mortality in Bangkok was statistically significant. Approximately 4000-5000 premature deaths each year in Bangkok metropolitan region are attributable to short term exposures to outdoor airborne PM (Radian International LLC, 1998). Studies investigating the impact of roadside PM (Tamura et al., 2003; Jinsart et al., 2002) found a significant direct relationship between levels of PM and exposure, and the prevalence of respiratory symptoms. This takes into consideration an annual ambient concentration of 64μg/m^3 affecting 5.7 million people resulting in 1092 excess deaths and 4550 cases of chronic bronchitis (World Bank, 2002).

The strongest effect of SO_2 and TSP on mortality was consistently seen for respiratory diseases in Beijing. Studies also revealed increased mortality associated with SO_2 pollution levels below the WHO 2000 air quality guideline value of 50 μg/m^3 (Xu et al., 1994, 1995 a and 1995 b; Wang et al., 1997; WHO 2000a; Chang et al., 2003). Kan et al., (2005) studied the relation between daily severe acute respiratory syndrome (SARS) mortality, ambient air pollution and other factors from 25 April to May 31, 2003 in Beijing. An increase of 10 mg/m^3 over a five day moving average of PM_{10}, SO_2 and NO_2 corresponded to 1.06, 0.74 and 1.22 relative risks of daily SARS mortality, respectively.

Similarly in Busan, SO_2 has been identified as a significant predictor for all cause deaths. An increase of 133$\mu g/m^3$ (50ppb) of SO_2 corresponded to a 3.6% increase in excess deaths (Lee et al., 2000).

A study by the National Building Research organization and the University of Colombo, Faculty of Medicine found a significant association between ambient air pollution (with respect to SO_2 and NOx) and acute childhood wheezing episodes in Colombo (Senanayake et al., 1999). According to the WHO the current level of PM_{10} in Colombo, which is approximately 80 $\mu g/m^3$, is sufficient to cause a 7% increase in daily mortality, 30-35% in bronchitis and other respiratory diseases (WHO, 2000a; Island The, 2003).

The high prevalence of allergy and asthma in Hanoi could be related to the degradation of air quality in the city. There are approximately 17 million motorcycles on Vietnam roads and an increasing number of motor vehicles. The health effect associated with mobile source pollution includes asthma, bronchitis, and premature deaths especially in children and elderly (Vietnam Panorama News Online, 2005).

In metro Manila an estimated additional 10,000 cases of acute bronchitis, 200 respiratory hospital admissions, 40 cardiovascular admissions, upto 200 cardiovascular deaths and upto 330 respiratory deaths were attributed to PM_{10} levels annually (Torres et al., 2004).

Medical records from hospitals in the Kathmandu valley revealed that the number of patients with acute respiratory infections (ARI) increased by 22.89% per year. Similarly the share of total ARI patients out of the total out patients department visits increased from 9.99 to 10.11 % from 1996 to 1998. Based on 1990 data, mortality in the Kathmandu valley due to air pollution is estimated to be 84 excess deaths annually, while the number of respiratory symptoms days were approximately 1.5 million (CEN, 2003).

A study in Taipei indicated that the higher levels of ambient pollutants (PM_{10}, NO_2, CO and O_3) increase the risk of hospital admissions for cardiovascular diseases especially during warm days-temperatures greater than or equal to 20degree (Chang et al., 2005). Another study observed a significant relationship between air pollution and daily mortality due to respiratory diseases in Taipei (Yang et al., 2004). Individual levels of SO_2, NO_x and NO_2 were positively correlated with the frequency of absence of primary school student due to illness while PM_{10} and O_3 did not show any correlation with absence frequency (Hwang et al., 2000). A study of the association between prevalence of asthma in middle school students, air pollution and weather confirms that asthma prevalence rates were highly correlated with certain traffic related air pollutants, CO and NO_x in Taiwan (Guo et al., 1999).

Shirazi and Harding (2001) have studied trends in ambient air quality and its possible impact on health in Tehran. The costs of deteriorating urban air quality in Kuala Lumpur and neighbouring Petaling Jaya are quantified by Moran et al.(2000) using contingent valuation method.

An evidence of respiratory health effects associated with air pollution comes from a survey of some of the major metropolitan areas in South Africa. This evidence indicated that domestic fuel combustion was associated with excess respiratory symptoms in young children (Von Schirnding et al., 1991). They reported that 29% of children under 5 years have experienced coughing and breathing problems in a two week recall period.

Apart from the outdoor pollution, indoor air pollution is a serious problem in the developing countries. The highest exposures, particularly among women, have been to indoor air pollution from the combustion of coal and biomass fuels for cooking and heating. Smith and Liu (1996) have reviewed the epidemiologic literature on indoor air pollution and lung cancer in the developing countries and found evidence of increased rates of lung cancer associated with indoor cooking and heating with coal mainly on China.

Saksena and Smith (2003) examine the potential impact of indoor air pollution on health in the developing countries, with a particular emphasis on exposure to particulates. Reviewing the evidence on the emissions, concentrations and populations exposed to indoor air pollution from traditional cooking fuels they found acute respiratory infection in children, and chronic obstructive lung disease, adverse pregnancy out comes and lung cancer in women.

The information so far gathered in this section shows that the people in developing countries are suffering from the pollution led diseases. The people of the developed countries are also suffering but the rate is comparatively low. The mortality and morbidity rates are also high in Asian developing countries as most of them are newly industrialized economies passing through transitional phase.

Here the study attempts to show the status on estimated total death caused by pollution led diseases and DALYs rate due to pollution in developing Asia[9].

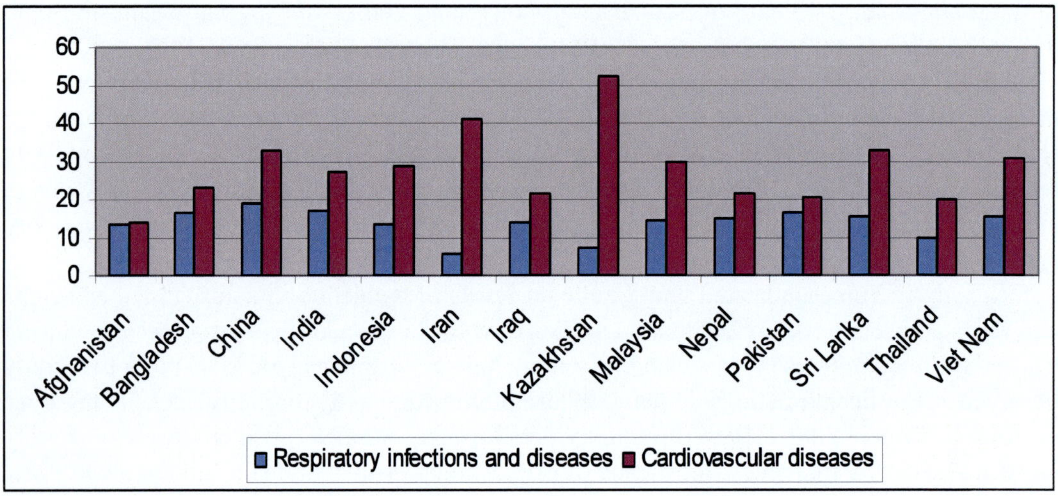

Source: World Health Organization, Department of Measurement and Health Information, December 2004

Figure 2.1. Estimated Total Deaths per thousand, by Cause in Asian Developing Countries, 2002.

The estimated total death per thousand is shown in figure 2.1. The percentage of deaths due to respiratory infections ranges between 6% to 19% (Appendix-A1). China is on the higher end, followed by India (Table 2.2, discussed in detail in next section), Bangladesh and Pakistan. The percentage of death due to cardio vascular diseases is really acute in Asian developing countries. The ranges are lying between 14% to 53%. In this case Kazakhstan is on the higher end followed by China, Malaysia, Indonesia, India and Vietnam.

[9] Since the current study is focusing on India, the death and DALYs rate due to pollution is confined only to developing Asian countries.

The estimated total DALYs ('000) is presented in figure 2.2. The estimated DALYs rate is lying between 7% to 13% (Appendix-A2) for respiratory diseases and 7-23% for cardio vascular related diseases. In DALYs count India has highest number of respiratory diseases compared to other Asian developing countries (Table2.3, discussed in detail in next section) followed by China, Pakistan, Afghanistan, and Bangladesh.

2.2. Impact of Air Pollution on Health in India

From the foregoing discussion it is evident that air pollution affects human health in a different way. In India there are various studies already done. Unfortunately most of them are scattered. Here we have tried to present a comprehensive picture as far as possible. But before going to literature it is better to have a glimpse of the status of health situation in India.

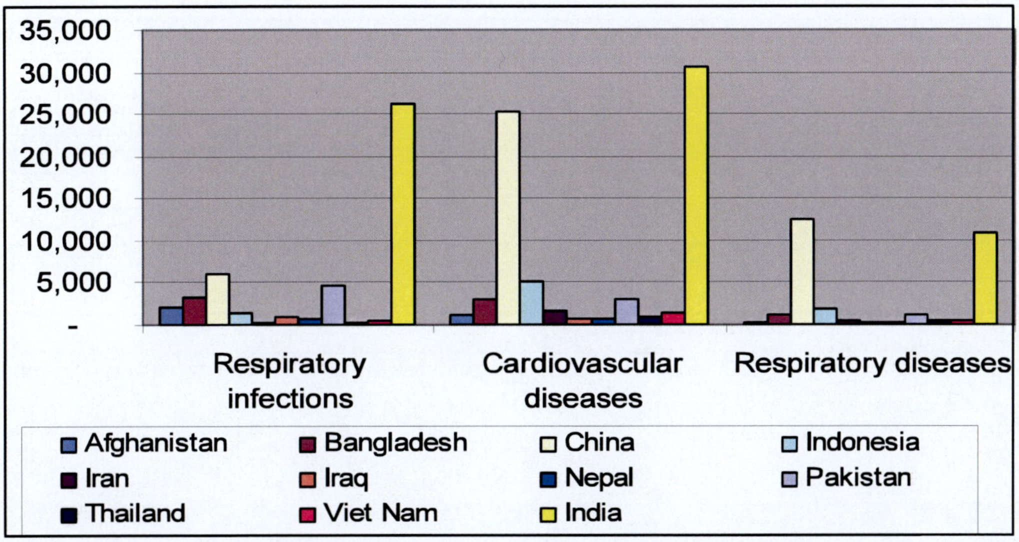

Source: World Health Organization, Department of Measurement and Health Information, December, 2004

Figure 2.2. Estimated Total DALYs ('000), by Cause in Asian Developing Countries, 2002.

Table 2.1 presents total deaths by cause. Out of all causes to death, 16% and 27% deaths are due to respiratory infections and respiratory diseases and cardiovascular problems in India respectively. But DALYs count (table 2.3) for the same are 13% and 10% respectively. Table 2.2 shows age specific death by cause. It reports which age groups are affected more due to pollution led diseases and how acute the diseases related to pollution are. The table shows that population belonging to the age group of 60+ are suffering more from Asthma and Bronchitis (82.9%), heart attack (60.7%) and Tuberculosis of lungs (32.4%) while children below 1 year are from pneumonia (54%). If we compare the past data of 1984, 1988 and 1998 (Mukhopadhyay and Forssell,2005), we observe that the respiratory problem and related diseases like asthma and bronchitis among the different age groups are increasing, especially the senior citizens are mostly affected.

Table 2.1. Estimated Total Deaths ('000), by Cause in India, 2002

Total Population ('000)		1,049,550
		Deaths
All Causes		10,378.50
Respiratory infections		1,123.10
1	Lower respiratory infections	1,107.90
2	Upper respiratory infections	14.7
3	Otitis media	0.5
Perinatal conditions		762.1
1	Low birth weight	542.5
2	Birth asphyxia and birth trauma	129.3
Cardiovascular diseases		2,810.00
1	Rheumatic heart disease	103.9
2	Hypertensive heart disease	49.7
3	Ischaemic heart disease	1,531.50
4	Cerebrovascular disease	771.1
5	Inflammatory heart diseases	57.8
Respiratory diseases		609.5
1	Chronic obstructive pulmonary disease	485.8
2	Asthma	57.1

Source: World Health Organization, Department of Measurement and Health Information, December 2004.

Tables 2.1 and 2.2 reveal that how fast the number of deaths is rising due to pollution. If we compare these figures with those of other developing countries in Asia (discussed in previous section), it is really discouraging. In DALYs count (table 2.3) India has highest number in case of respiratory diseases compared to other Asian developing countries. The DALYs and death count data show that the situation in India is acute compared to other developing countries in Asia. A study by the World Bank group (World Bank ,2001) on premature mortality and burden of diseases due to air pollution in India and other regions of the world shows that the premature deaths will be 950,000 p.a. for India, and the burden of diseases will be 27.1 million DAILYs p.a. in the year 2020. These figures confirm our concern about the severity of the impact of air pollution on human health of the future generation of India.

We have seen in Chapter 1, how the emission problem is growing in India. In this section, we will piece together the literature on air pollution and health only from the Indian metros.

Studies involving air pollution and its impact on health though few in India, are available on metro cities. Studies by Agarwal et al. (1999), Sikdar and Mondal (1999), Chitkara (1997), Sinha and Bandyopadhyay (1998), Romieu and Avila (2003), Saksena and Smith (2003) analyse the causal relation between air pollution and health hazards in India.

Table 2.2. Percentage Distribution of Deaths due to Selected Killer Diseases from Air Pollution By Age Groups, India, 1998

Sl.	Diseases	Age Groups In Years								
No.		Below 1 Yr.	1-4	5-14	15-24	25-34	35-44	45-59	60+	Total reported deaths (no.)
1.	Asthma and bronchitis	0.8	0.8	0.7	0.8	1.2	2.1	10.7	82.9	5055
2.	Heart attack	0.1	0.2	0.8	1.9	3.4	7.9	25.1	60.7	3577
3.	Pneumonia	54.0	23.7	7.6	2.0	1.2	1.4	2.5	7.6	1282
4.	T.b. of lungs	0.6	1.2	1.8	6.2	13.1	18.5	26.2	32.4	1987
5.	Prematurity	100.0	-	-	-	-	-	-	-	1762

Source: Survey of Causes of Death - 1998; Registrar General Of India.

Table 2.3. Estimated Total DALYs ('000), by Cause in India (Global Burden of Disease), 2002

Total Population ('000)		1,049,550
		DALYs rate
All Causes		299,910
Respiratory infections		26,094
1	Lower respiratory infections	25,556
2	Upper respiratory infections	258
3	Otitis media	280
Cardiovascular diseases		30,481
1	Rheumatic heart disease	1,986
2	Hypertensive heart disease	619
3	Ischaemic heart disease	15,316
4	Cerebrovascular disease	7,357
5	Inflammatory heart diseases	1,153
Respiratory diseases		10,789
1	Chronic obstructive pulmonary disease	5,787
2	Asthma	3,071

Source: World Health Organization, Department of Measurement and Health Information, December 2004.

A couple of studies focused on Mumbai. Since 1990s World Bank is engaged to estimate the impact of pollution, its abatement cost and health. A study on Mumbai shows that annual average TSP concentration has increased about 50 percent from 1981 to 1990, to reach 270$\mu g/m^3$. World Health Organization and national guidelines for PM_{10} frequently and substantially exceeded in Mumbai. 97 percent of the population lives in areas where the WHO air quality guideline for particulate exceeded. Studies point to the resulting health effects—more cases of colds, chronic bronchitis, asthma and general decline in lung function.

Using dose response equations developed in the United States, this report estimates that air pollution causes 2,800 cases of excess mortality, 60 million respiratory symptom days, and 19 million restricted activity days, at a total cost of Rs.18 billion per year(World Bank, 1997a).

Several epidemiological studies have shown that in Mumbai, with moderately raised pollution levels (SO_2, NO_2 and TSP), there was an increased occurrence of dyspnoea, chronic and intermittent cough, frequent colds, chronic bronchitis and cardiac disorders, mainly cough, high blood pressure and deaths due to non tuberculous respiratory and ischaemic heart disease (Kamat,2000).

Some studies concentrate only on Delhi. Cropper et al. (1997) have examined the impact of particulate air pollution (SO_2, TSP) on daily mortality in Delhi. They have found a positive significant relationship between particulate pollution and daily non-traumatic deaths, as well as deaths from certain causes (respiratory and cardiovascular problems) and for certain age groups. The average TSP level in Delhi was 378 $\mu g/m^3$ approximately five times higher than the WHO annual average standard. The particulates have the greatest impact in the 15-44 age groups (over 74%). On an average, a 100 $\mu g/m^3$ increase in TSP is associated with 2.3 % increase in deaths in Delhi.

A study conducted by the all India Institute of Medical Sciences in New Delhi showed that exposure to PM has an impact on respiratory morbidity. Results indicated that most common symptoms related to air pollution were irritation of eyes, cough, pharyngitis, dysponea and nausea. The increase in hospital admissions due to respiratory morbidity has also been correlated with a rise in PM levels (Mashelkar et al., 2002).

Several studies are also available for other major cities in India.

In Bangalore, respiratory problems among children have risen threefold during the last 20 years (Paramesh et al., 2001). The incidence of respiratory ailments such as asthma during 1979, was only 9% in the children population of the district. By 1999, it had risen to 29.5%. Corresponding increase in the number of industries and automobiles was also witnessed[10].

An association between ambient air pollutants and respiratory symptoms complex (RSC) in preschool children, a cohort of 664 children between the ages of 1 month to 4.5 years, was found at Lucknow (Awasthi et al., 1996). Exposure to ambient air sulphur dioxide, oxides of nitrogen and SPM on the day of the interview or in the week prior to it, was assessed by ambient air monitoring at nine centres in the city. The cumulative incidence of RSC was observed to be 1.06 and the incidence density per 100 days of follow-up was 1.63.

A cross-sectional study was performed in one industrial and one non-industrial town in Punjab State, northern India(Kumar et al., 2004). Ambient air quality samples were collected and analyzed each week for 2 years. Total 3,603 individuals greater than 15 yr old were interviewed and whose lung functions were measured spirometrically. Their biomarkers were categorized in terms of obstructive or restrictive defects. Levels of total suspended particulates, nitrogen oxides, sulfur oxides, carbon monoxide, and ozone were significantly higher in the study town than in the reference town. The prevalence of chronic respiratory symptoms (cough, phlegm, breathlessness, or wheezing) was 27.9% and 20.3% in the study and reference towns, respectively. That of obstructive ventilatory defect was 24.9% and 11.8%, respectively. Logistic regression analysis showed that residence in the study town was

[10] Globally, over 180,000 people die from asthma each year. India has approximately 15–20 million asthmatics (Narain,2002).

independently associated with chronic respiratory symptoms and spirometric ventilatory defect after controlling for other demographic effects.

Some of the studies covered few metros in India or the country as a whole. Among them a study in 1995 (Agarwal et al., 1999) has formulated a model to establish the relationship between air pollution and human morbidity and mortality. The model was subsequently used to assess environmental and health condition in India. Using air quality data for 1991-92 provided by Central Pollution Control Board (CPCB) from 290 monitoring stations in 92 Indian cities and towns researchers have found that air pollution results in 40351 premature deaths in India. Calcutta, Mumbai and Delhi accounted for 5726 (14%), 4477 (11%), and 7491 (19%), respectively. Substituting the CPCB air quality data for 1991-92 by the corresponding 1995 data, researchers at the Centre for Science and Environment (CSE) have found that the figures for number of premature deaths increased to 51779 – a rise of 28%. Calcutta, Delhi, Mumbai, Kanpur and Ahmedabad accounted for 10647,9859,3639,3006 premature deaths respectively (Brandon and Hommann, 1991-92 and Anon, 1997). These cities account for 66% of the total premature deaths from air pollution in India. The total estimates of annual episodes of illness due to SPM in the air have increased from 19805388 in 1991-92 to 25645721 in 1995. Thus the country is paying a heavy price as a result of deaths and incidences of illness due to ambient SPM.

Chitkara (1997), on the other hand presents a brief review on air pollutants and related health hazards. According to this study air pollution depends on three factors : 1) source of emissions 2) meteorological conditions and 3) time. It explains the factors affecting air pollution, emissions discharges and their sources (vehicular emission, domestic emission, industrial emission, emission due to energy).

TERI (1997) has carried out few estimates based on the effects of SO2, particulate matter, carbon monooxide and carboxyhaemoglobin at various concentrations (ppm), exposure (time) and corresponding health effects in India. A study by Sinha and Bandyopadhyay (1998) has tried to capture the metallic constituents of aerosol present in biosphere, which have been identified as potential health hazards to human beings. They have examined the concentration of Cd (cadmium), Zn (zinc), Fe (iron), Pb (lead) and Cr (chromium) in ambient air of Delhi, Mumbai, Calcutta and Chennai cities in India. Those pollutants are insidious because non bio degradable in nature. They attempt to assess the present level of trace elements in the above major cities or towns, potential sources and health impacts. They recognize that trace elements emanate from industrial, commercial and vehicular activities and create potential health hazards for living beings. They have concluded that controlled emission from industrial operations would help to keep the metallic concentration within limits in the ambient air. A similar study by Sharma et al. (1999) concentrates on acute respiratory infections in urban slum area in India.

Chabra (1996) establishes air pollution as one of the major causative factors behind the 10-12% incidence of bronchial asthma in the age group of 5-16 years in India. And it is increasing significantly due to the rapid growth of transport sector. Chabra also adds that chronic and complicated cases such as patients with lung ruptures, fibrosis and pneumonia are due to pollution, especially traffic pollution. While reporting 40,000 premature deaths due to air pollution in Indian cities in 1991-92, a World Bank study had categorically stated that SPM and SO_2 are responsible for over 95% of health damage, and the condition of asthma sufferers in Indian metropolitan cities is the worst. PM_{10}, NO_X, SO_2 and Ozone worsened the condition of asthmatics. They experience symptoms like shortness of breath, wheezing, and coughing and sometimes even death.

Mukhopadhyay and Forssell (2005) have estimated air pollution (CO_2, SO_2, and NOx) from fossil fuel combustion in India. Input–Output Structural Decomposition Analysis approach is used to find out the sources of changes. They also estimate the emissions of CO_2, SO_2 and NO_x for the year 2001–2002 and 2006–2007. A link between emission of pollutants and their impact on human health is also analysed. The main factors for emission changes are the pollution intensity, technology, and the volume of final demand. These results are quite different to those observed in some other studies. They conclude that a large number of people are affected by respiratory infection especially lower respiratory infections. The pollutants like SO2 and NOX have caused these respiratory infections. The study found a close linear relationship between pollution and health impacts.

In pollution and health literature a few studies focused on Calcutta. Some of them have been captured here.

In 1995 an estimated 10,647 premature deaths were attributed to air pollution in Kolkata (Ghose, 2002; Schwela et al.,2006). Studies have demonstrated that children inhaling polluted air in Kolkata suffer from adverse lung reactions and generic abnormality in exposed lung tissues (Lahiri et al., 2000). Approximately 47% of Calcutta's population suffers from lower respiratory tract symptoms with the lungs of city residents being approximately seven times more burdened compared to their rural counterparts due to air pollution (Roy et al.,2001; WBPCB, 2003; Schwela et al.,2006). Sikdar and Mondal (1999) have conducted a study on air pollution (TSP, SO_2, NOx, CO_2) in Calcutta. They have identified vehicular emission as a primary source of these pollutants. They have suggested that an air quality management on reducing stationary source and mobile source emissions will help to mitigate the air pollution and improve the quality of life.

Chabra et al. (2001) showed that an increase in the economic status of the individual significantly decreased the prevalence of respiratory symptoms among those who were exposed long term to high levels of air pollution. Evidence suggests that an asthma sufferer exercising in the presence 290 $\mu g/m^3$ SO_2 will show severe symptoms of asthma within minutes. In Calcutta, SO_2 levels are getting dangerously high – in some residential areas SO_2 levels have touched 175 $\mu g/m^3$ during winters. Calcutta is placed sixth among the 41 most polluted cities when it comes to SO_2 and SPM levels, according to Global Pollution and Health, a report published in 1996 by WHO and the United Nations Environmental Programme (UNEP). The annual average level of SPM by WHO standards is 60-90 $\mu g/m^3$ and Calcutta's average was 344.3 $\mu g/m^3$ in 1995. This increases more in winters. Over the past decade, there has been a significant increase in the incidence of lung diseases, asthma and bronchitis and respiratory and cardio-vascular problems in Calcutta. The experts point out the auto emission, which is doubling in every decade, to be one of the most important pollutants causing vehicular pollution. Calcutta has a density of 814.80 vehicles, the highest among per km road length as compared to 766.31 for Mumbai, followed by 616.58 Chennai and 170 Delhi (Calcutta News, 2005).

2.3. VALUATION OF HEALTH COST

It is difficult to estimate the health costs in the developing world. There are few studies which already attempted to estimate the valuation of health like life years lost, loss of working days, as well as loss of salary.

Estimates of economic health cost in India have been made by experts. Cropper et al. (1997), Chabra (1996), Brandon and Homman (1991-92), Anon (1997), Murty, Gulati and Mukherjee (2002) have tried to estimate the health cost from air pollution in India. Experts have estimated that economic health costs due to growing levels of SPM, SO2 and NOX range between Rs. 1755crore and Rs. 7252 crores. Similarly estimates of monetary losses due to sickness caused by high levels of SPM are between Rs. 107 crores and Rs. 213 crores in Delhi (Brandon and Homman, 1991-92 and Anon, 1997). National Institute of Public Finance and Policy, New Delhi was commissioned by the expert committee on auto fuel policy to investigate the costs of health damages from vehicular emissions. The results showed that the annual health damage of pre euro emissions of 25 Indian cities ranged from US$14 million (Rs 6.8 billion) to US $191.6 (Rs 93.1 billion) (Mashelkar et al.,2002). According to Mohanraj(2004) the health costs due to air pollution in India is alarming and it ranges between 517 -2102 million USD.

Lvovsky (1998) finds that the costs to society, part of which is direct productivity loss, due to air pollution in largest India cities are as high as nearly one-tenth of the income generated in these cities from all economic activities. This analysis clearly shows that India suffers from a disproportionally heavy health burden of urban air pollution by international comparison.

Another study in this direction has been made by Gupta (2006) on reductions in air pollution in the urban industrial city of Kanpur in India. This study estimates the monetary benefits to individuals from health damages avoided as a result of reduction in air pollution. For measuring monetary benefits, the study considers two major components of health cost — the loss in wages due to workdays lost and the expenditure incurred on mitigating activities. The study estimates that a representative individual from Kanpur would gain Rs 165 per year if air pollution was reduced to a safe level. The extrapolated annual benefits for the entire population in the city will be Rs 213 million.

Murty, Gulati and Mukherjee (2002) have estimated benefits from controlling urban air pollution in the metropolitan areas of Delhi and Calcutta in India using hedonic property price method and the primary data collected from the household surveys. Estimates of hedonic property price function and the implicit marginal price or the household's marginal willingness to pay function for air quality are obtained first separately for each city and then by using the pooled data for both the cities. Also the study provides estimates of aggregate consumer surplus benefits to the households of each city from reducing the air pollution concentration of suspended particulate matter (SPM) from the current levels to the safe WHO or MINAS standards for India. They conclude that the welfare gains from reducing air pollution from the current levels to the safe levels in the cities of Delhi and Calcutta as revealed through the location choices of houses by the households are very high. A representative household gets an annual benefit of Rs.19870.70 in Delhi and Rs.84355.71 in Calcutta. When the benefits are extrapolated to all the urban households in each city, the households in Delhi get benefits worth of Rs.46655.2 million while those in Calcutta get Rs.26635.3 million. These benefits though look high, are not in comparison to the cost the Government and polluters have to bear to reduce the air pollution levels from the current level to the safe level. In fact these benefit estimates justify the cost of environmental policy changes like introducing CNG operated vehicles, and substituting the metro rail to road transport and the relocation of polluting industries in the cities.

The cost of health impacts in Mumbai has been estimated to be total of US$44.9 million for costs attributed to mortality using the "loss in salary approach" and US$206.3 million for costs attributed to morbidity (Srivastava and Kumar,2002). Another study in Mumbai found high health costs of approximately 19% of the income of patients suffering from severe air pollution related attacks (Parikh and Hadkar, 2003).

Estimates are also available for other developing Asia in this respect. Jakarta is one of the most polluted cities in the world. Air pollution in Jakarta is above the safe limits specified by the World Health Organization. It is estimated that the health cost of Jakarta's air pollution in 1999 reached $US220 million (Resosudarmo and Napitapulu, 2004). The new vehicle emission standard policy is more effective in reducing the health costs associated with air pollution. They show that the total health cost reduction in 2015 that can be achieved by implementing all policies is approximately 428 billion Rupiah (or $US402 million), 0.25 per cent of Jakarta's GDP in 2015.

Cost of illness and willingness to pay estimates of the damages from minor respiratory symptoms associated with air pollution are compared using data in Taiwan in 1991-92 (Alberini and Krupnick ,2000). A contingent valuation survey is conducted to estimate WTP to avoid minor respiratory illnesses.

The annual cost of morbidity resulting from PM_{10} is estimated to be US $6.1 million and total health damage to be US $7.2 million in Kathmandu (CEN, 2003; Schwela et al.,2006).

In Philippines the impacts and costs on health attributed to air pollution particularly PM_{10}, were estimated at US$392 million for 2001, based on the monetary costs for excess deaths(1915), treatment of chronic bronchitis incidence (8439) and respiratory symptom incidence (50.5 million) linked to PM_{10} (World Bank 2001).

A time series study has been conducted on the impact of particulate air pollution on daily mortality. The total social costs resulting from air pollution in Busan are approximately US$2.5 billion. The costs of PM_{10} account for the largest share at US$880 million (Schwela et al.,2006).

Air pollution is responsible for an estimated 6000 deaths in Dhaka city each year and on average each city dweller spends approximately US$12 per year on medical treatment for pollution related illness (Islam, 2003; Schwela et al.,2006).

The estimated health impacts and costs of PM10 in Bangkok for 2000 amount to $424 million (Schwela et al.,2006).

Karimzadegan et al.,(2008) show that the total daily health damage costs of the air pollution in Great Tehran area has been estimated at 16224 US$ per each unit increase of PM_{10}, 28816 US$ per each unit increase of CO, 1927 US$ per each unit increase of NO_2 and 7739 US$ per each unit increase of SO_2.

The above review of the literature establishes the fact that diseases such as asthma, bronchitis, and respiratory kind of problem are due to air pollution. Further, the literature presents the country as a whole or region wise assessment. But hardly any study focuses on the contribution of different income groups on the generation of emission, factors responsible and finally to evaluate the impact on health. India being a developing country with rather high air pollution and acute poverty should address this problem seriously. The current study is to make an attempt in that direction.

Chapter 3

MODEL FORMULATION

In this chapter the present study formulates the model based on Input-Output Structural Decomposition analysis. Two models have been developed. Model 1 describes the estimation of pollutants emission (CO_2, SO_2 and NOx)[11] and factors responsible for changes in emission. Model 2, as an extension of model 1, incorporates different income groups.

ADVANTAGE OF THE FRAMEWORK

The methodology for relating economic activity to the natural environment in an Input-Output framework is convenient and popular. In particular, when modeling the use of various fuels by an economy with Input-Output method, the distinction must be made between fuels used by industries, to allow the production of the total output of those industries, and fuels used directly by final consumers. It is well known that Input-Output analysis is a suitable tool for assessing resource or pollutant embodiments in goods and services. Another advantage lies in assessing the role of different income groups through multiplier effect. So it is a unique technique to address the objective of our study.

MODEL 1

The model starts with the basic concepts of Input-Output framework of Leontief, (1951). Mathematically, the structure of the input-output model can be expressed as:

$$X = Ax + Y \qquad (1)$$

The solution of (1) gives

$$X = (I - A)^{-1} Y \qquad (2)$$

[11] In this study we consider, CO_2, SO_2 and NO_x among all other air pollutants. We used IPCC guideline to estimate the total emissions in India. This guideline is not applicable to other pollutants like SPM, PM_{10}, $PM_{2.5}$ and so on. Sulphur and nitrogen oxides are dangerous to human health. CO_2 does not have direct impact on human health, but a significant relationship exists between carbon monoxide and daily mortality. Diseases due to these pollutants are described in Chapter 1.

where $(I - A)^{-1}$ is the matrix of total input requirements .For an energy input-output model, the monetary flows in the energy rows in equation (2) are replaced with the physical flows of energy to construct the energy flows accounting identity, which conforms to the energy balance condition (Miller and Blair, 1985). We apply a "hybrid method" based on Miller and Blair (1985), and it always conforms to energy conservation conditions.

In equation (1) and (2), X is a hybrid unit total output vector (nx1) in which the output of energy sectors is measured in million tonnes of coal replacement (mtoe), while the output of other sectors is measured in million rupees (M.RS). Y is a hybrid unit final demand vector (nx1), in which the final demands for different types of energy are measured in mtoe, while the final demands for the outputs of other sectors are measured in M.RS. A is a hybrid unit technical coefficient matrix (nxn), in which the unit of the input coefficients of energy sectors from energy sectors is mtoe / mtoe; the unit of the input coefficients of energy sector from non energy sectors is M.RS/ mtoe; the unit of the input coefficients of non energy sectors from energy sectors is mtoe /M.RS; and the unit of the input coefficients of non-energy sectors from non-energy sectors is M.RS/M.RS. I is an identity matrix (nxn).

For estimating the carbon, sulphur and nitrogen, we followed the IPCC guidelines[12]. A detail estimation procedure is described below. We have computed the amount of CO_2, SO_2 and NO_x emission that takes place in the production of various activity levels by extending the conventional input-output framework. We apply the fuel specific carbon, sulphur and nitrogen emission factors[13] to the row vector of fossil fuel sector of the respective Input-Output table to estimate the total CO_2, SO_2 and NO_x emitted by coal and oil sector.

On the basis of the procedure mentioned in note we calculate the direct carbon dioxide, sulphur dioxide and nitrogen oxide emission coefficient and total (direct and indirect) carbon dioxide, sulphur dioxide, and nitrogen oxide emission coefficient.

Let C = C (j) (**)

It is a vector of fossil fuel emission coefficients representing the volume of CO_2, SO_2 and NOx emissions per unit of output in different sectors. That is when the sectoral volume of CO_2, SO_2 and NOx emission is divided by sectoral output then it gives us the direct CO_2, SO_2 and NOx emission coefficient. The direct and indirect carbon sulphur and nitrogen emission coefficient of sector j can be defined as $C_j r_{ij}$, where r_{ij} is the (i,j)th element of the matrix $(I-A)^{-1}$. The direct and indirect CO_2, SO_2 and NO_x of a sector is defined as emission caused by the production vector needed to support final demand in that sector. This would depend not only on the direct and indirect emission coefficient of that sector but also on the level of sectoral final demand.

12 IPCC guideline = energy consumption*emission factor*molecular weight ratio*fraction of carbon sulfur and nitrogen oxidized.

13 We follow the normal convention of measurement of carbon, sulfur and nitrogen dioxide in carbon sulfur and nitrogen equivalent units. It gives the total quantity of CO_2, SO_2, and NO_x emitted owing to burning of fossil fuel (coal, oil) inputs used by various production sectors.

	Emission factor for Coal	Emission factor for Oil	Molecular weight ratio	Fraction of carbon oxidised
Carbon	0.55 (mt of CO_2) /mt	0.79(mt of CO_2)/mt	3.66	0.98
Sulfur	.003 (mt of SO_2) /mt	0.015 (mt of SO_2)/mt	2	0.97
Nitrogen	.018(mt of NO_x.) /mt	0.001 (mt of NO_x)/mt	3.28	0.96

I) Emission Model

Now in equation form of CO_2, SO_2 and NO_x emissions from fossil fuel combustion can be calculated from fuel data in the following manner.

$$F = C_tL_1X = C_t L_1 (I - A)^{-1} Y \qquad (3)$$

Here F as a vector, giving the total quantity of CO_2, SO_2 and NO_x emissions from fossil fuel combustion only.

C as a vector of dimension m (mx1), of coefficients for CO_2, SO_2 and NO_x emissions per unit of fossil fuel burnt.

L_1 as a matrix (mxn) of the industrial consumption in energy units of m types of fuel per unit of total output of n industries.

Subscript t denotes the transpose of this vector.

In equation (3) C_tL_1= S carries only direct requirement of CO_2 , SO_2 and NO_x intensities from industries and $C_t L_1(I - A)^{-1}$ gives the direct as well as indirect requirement of CO_2, SO_2 and NO_x intensities from industries .

So equation (3) explains the CO_2, SO_2 and NO_x emissions due to fossil fuel combustion from production activities.

II) Structural Decomposition Analysis

Next, we develop a Structural Decomposition Analysis [SDA] for this model to estimate the changes in emission in each period as well as to capture the factors responsible for such changes in emission. It is a technique to study over period changes. It has become a major tool for disentangling the growth in some variables over time, separating the changes in the variable into its constituent parts. SDA[14] seeks to distinguish major sources of change in the structure of the economy broadly defined by means of a set of comparative static changes in key parameters of an Input-Output table.

The total industrial CO_2, SO_2 and NO_x emissions (TE) can be expressed as (Mukhopadhyay and Forssell, 2005):

$$TE = \Delta F = SRY = S (I - A)^{-1} Y \qquad (4)$$

where $R = (I - A)^{-1}$

Here S represents the industrial CO_2, SO_2 and NO_x intensity.

[14] A number of studies concentrate on energy and environment (greenhouse gas emissions) by applying Input-Output and SDA. Works in this area by Rose and Chen (1991) for Taiwan; Lin and Polenske (1995); and Lin (1998) for China; Lin and Chang (1996), Chang and Lin (1998) for Taiwan Province of China; Gay and Proops (1993) for the UK; Breuil (1992) for France; Liakas et al., (2000) and Hann (2001) for the European Union countries and the Netherlands, respectively; Common and Salma (1992) for Australia; Bossier and Rous (1992) for Belgium; Hayami et al., (1993), Han and Lakshmanan (1994) for Japan; Wier (1998) for Denmark; Zhang and Folmer (1998) for Germany; Maenpaa (1998) for Finland, Mukhopadhyay and Chakraborty (1999,); Mukhopadhyay (2002 and 2002a); Mukhopadhyay and Forssell (2005) for India should be mentioned.

According to the structural decomposition analysis method, the change in total CO_2, SO_2 and NO_x emissions between any two years i.e. year o and year t can be identified as:

$$TE = \Delta F = S_t (I - A_t)^{-1} Y_t - S_o (I - A_o)^{-1} Y_o \quad (5)$$

$$= S_t R_t Y_t - So\,Ro\,Y_o \quad (6)$$

$$= S_t R_t Y_t - S_o R_t Y_t + S_o R_t Y_t - S_o R_o Y_o \quad (7)$$

$$= \Delta S R_t Y_t + S_o R_t Y_t - S_o R_o Y_o \quad (8)$$

$$= \Delta S R_t Y_t + S_o R_t Y_t - S_o R_o Y_{t-} + S_o R_o Y_t - S_o R_o Y_o \quad (9)$$

$$= \Delta S R_t Y_t + S_o \Delta R Y_t + S_o R_o Y_t - So\,R_o Y_o \quad (10)$$

$$= \Delta S R_t Y_t + S_o \Delta R Y_t + S_o R_o \Delta Y \quad (11)$$

The first term of equation (11) reflects the CO_2, SO_2 and NO_x emission changes due to the changes of CO_2, SO_2 and NO_x intensity of various industries.

The second term of Equation (11) defines the CO_2, SO_2 and NO_x emission changes due to the changes in technical coefficient matrix.

And the third term of Equation (11) refers to the CO_2, SO_2 and NO_x emission changes due to the changes in the final demand of various industries.

Here t refers to the current period and o to the previous period.

Only fuel NO_x has been considered. Thermal NO_x has not been taken into consideration.

MODEL 2

Extension of Model 1

To calculate the contribution of the above emission changes contributed by the different income groups[15] the previous model has been extended accordingly. The final demand vector Y (from equation 4) has been treated separately by breaking the total final demand as

$$Y = Y_1 + Y_2$$

$$\text{where, } Y_1 = C_l + C_m + C_h \quad (12)$$

$$Y_2 = Y_2 \quad (13)$$

The terms C_l , C_m and C_h indicate the vector of household consumption belonging to lower, middle and higher income groups respectively.

15 income groups/classes considered here as different expenditure class

The term Y_2 signifies the vector of other final demand components like government consumption, change in stock, investment, export and import.

Now if we introduce equation (12) and (13) into equation (11) then it finally formed as

$$=\Delta S\, R_t (C_{lt} + C_{mt} + C_{ht} + Y_{2t}) + S_o\, \Delta R\, (C_{lt} + C_{mt} + C_{ht} + Y_{2t}) + S_o\, R_o\, \Delta (C_l + C_m + C_h + Y_2) \quad (14)$$

So the first term of equation (14) reflects the changes in intensity of CO_2, SO_2 and NOx term by considering the different final demand groups presented in equation (14a).

$$=\Delta S\, R_t\, C_{lt} + \Delta S\, R_t\, C_{mt} + \Delta S\, R_t\, C_{ht} + \Delta S\, R_t\, Y_{2t} \quad (14a)$$

Likewise, the second term of equation (14) covers the changes in technical coefficient of CO_2, SO_2 and NOx term by considering the different final demand groups explained in equation (14b).

$$= S_o\, \Delta R\, C_{lt} + S_o\, \Delta R\, C_{mt} + S_o\, \Delta R\, C_{ht} + S_o\, \Delta R Y_{2t} \quad (14b)$$

Finally, the third term of equation (14) reflects the changes in final demand of CO_2, SO_2 and NOx term by considering the different final demand groups described in equation (14c).

$$= S_o\, R_o\, \Delta\, C_l + S_o\, R_o\, \Delta\, C_m + S_o\, R_o\, \Delta\, C_h + S_o\, R_o\, \Delta\, Y_2 \quad (14c)$$

Each income group specific and rest of the final demand's contribution for CO_2, SO_2 and NOx emissions can be figured out from equations (14a, 14b and 14c). By this categorisation we can estimate the degree of responsiveness of the factors responsible for all emissions among each income group in a special form. It is presented in equations 15a through 15d.

$$L=\Delta S\, R_t\, Clt + S_o\, \Delta R\, Clt + S_o\, R_o\, \Delta\, Cl \quad (15a)$$

$$M=\Delta S\, R_t\, Cmt + S_o\, \Delta R\, Cmt + S_o\, R_o\, \Delta\, Cm \quad (15b)$$

$$H= \Delta S\, R_t\, Cht + S_o\, \Delta R\, Cht + S_o\, R_o\, \Delta\, Ch \quad (15c)$$

$$Y2= \Delta S\, R_t\, Y2t + S_o\, \Delta R Y2t + S_o\, R_o\, \Delta\, Y2 \quad (15d)$$

Equations 15a, 15b, 15c, and 15d combine the effect of factors responsible for each income group.

Chapter 4

DATA SOURCE AND PROCESSING

To implement the model and to conduct the Structural Decomposition Analysis of energy consumption changes we require Input-Output data, price indices, energy flow data and emission data (IPCC guideline). Input-Output tables of the Indian economy for the years 1983-84, 1989-90, 1993-94, 1998-99, 2003-4 prepared by CSO (1990, 1997, 2000, 2005, and 2008) are used. For price deflator National Accounts Statistics (CSO, Various issues) and for energy data Energy Statistics Report (2006) published by CSO have been used.

Consumption Expenditure of different commodities for different expenditure classes have been collected from the disaggregated data of various rounds of National Sample Survey (38th, 45th, 50th, 55th and 60th round NSSO, Government of India).

AGGREGATION OF INPUT-OUTPUT TABLE

Input-Output tables are Commodity by Commodity tables consisting of 115x115 sectors for the year 1983-84, 1989-90, 1993-94, 1998-99 and 130x130 for the year 2003-4. These sectors have been aggregated to 47 sectors on the basis of the nature of commodities and energy intensiveness. We have considered three energy sectors coal, crude oil and natural gas, and electricity separated and rest of the non energy sectors have been aggregated to 44 sectors. The aggregation scheme is given in Appendix B1 and B2.

PRICE INDICES

To make the Input-Output tables of 1983-84, 1989-90, 1993-94, 1998-99 and 2003-4 comparable the tables must be evaluated at some constant prices. We use 1993-94 as a base year and adjust 1989-90, 1993-94, 1998-99 and 2003-4 table to 1993-94 prices using suitable price indices available from National Accounts Statistics.

ENERGY DATA

We convert the monetary units of energy sectors into physical unit from the energy data published by CSO, Energy Statistics Report, 2006. Three energy sectors like coal as million tonnes, crude petroleum in million tonnes, natural gas in million cubic meter, and electricity in T.W.H have been converted to million tonne oil equivalent(mtoe).

NSSO

a) Data Specification

Household Consumer Expenditure in India, NSSO data for 38th Round (1983-84), 45th Round (1989-90), 50th Round (1993-94), 55th Round (1999-2000)[16] have been collected from NSS, New Delhi which were in the .dat format, and converted to the required format using SPSS 10.0 while data for 60th Round (2004) have been downloaded from the CSO website.

b) Conversion of NSS data in compatible form

NSS data are compatible to many data bases and econometric software. But in our case the consumer expenditure dataset is huge. In case of 1993-94 data for food item consist of around 26 lacs rows for urban and 37 lacs for rural, for other items are in figure of 14 lacs and 13 lacs for rural and 10lacs and 11 lacs for urban respectively. The data of 1999-2000 consist of around 10million sample for rural and urban. Thus we have used SPSS 10.0 to read and analyze data in the required format. The 30,000 sample households of rural and urban have been surveyed for the 60th round for the year 2003-4.

c) Nature and components

Disaggregated raw data have been arranged in terms of several variables like item code, sector code, state code, household's no, total expenditure, and monthly per capita expenditure(MPCE) etc.

d) Data Analysis using SPSS 10

NSSO 38th, 45th , 50th and 55th Round data (for the year 1983-84, 1989-90,1993-94 and 1999-2000 respectively) have been arranged in terms of item code, expenditure on those items and then MPCE (item codes for different blocks and the required block levels for our purpose are extracted from the huge data set). But the 60th round for the year 2003-4 is easier to deal with because of small sample set compared to other round's data.

Data are arranged and sorted according to different expenditure class, which are further segregated in terms of different expenditure class like MPCE for lower income group (LIG), middle-income group (MIG), higher income group (HIG). This classification has been made for the year 1993-94. Due to changes in prices, the size of the income group will change for

[16] Though the current study considered Input-Output table of 1998-99, the data coverage of NSS in the year 1998-99 is not sufficient to carry out study. So the year 1999-2000 data of NSS has been used. The year 1999-2000 has been chosen due to its extensive data coverage more than 10 million.

the year 1983-84, 1989-90, 1999-2000 and 2003-4. Consumption expenditure for the year 1983-84, 1989-90, 1999-2000 and 2003-4 are available at current prices so necessary price adjustments have been made. Necessary deflators have been used to make all the income groups (for the year 1983-84, 1989-90, 1999-2000 and 2003-4) at the price 1993-94. So the income groups classifications are at 1993-94 prices are as follows:

Rs 0-6000 is classified as lower income groups; middle income groups leveled as Rs. 6000-12,000, and Rs. 12,000 and above belong to upper income groups.

Thus we have computed per capita expenditure data from the available dataset and then we segregated the expenditure data according to the three income groups.

Disaggregated data at all India level consists of 700 items for the year 1983-84, 1989-90, 1999-2000 while for 2003-4, the number of items were around 40 but it is prepared from the originally 700 items. The detail consumer expenditure item list is presented in Appendix B3.

We aggregated item-wise consumer expenditure data of different income groups of NSSO data on the basis of input-output classification table for the respective years. Aggregation scheme is enclosed in Appendix B1 and B2.

The percentage share of each income group on the sector wise expenditure has been calculated for the respective years and is displayed in Appendix C1-C5. These shares are used to allocate the sectoral private consumption expenditure recorded in Input-Output table among three income groups.

Chapter 5

MODEL ESTIMATION AND ANALYSIS OF RESULTS

In this chapter the study analyses the results based on model 1 and model 2 described in chapter 3. Before discussing the results let us try to capture the Indian economy and its pattern of energy consumption, because emissions and sources of emissions are mainly due to this. During last 50 years after independence, the demand for energy, particularly for commercial energy, registered a high rate of growth (commercial primary energy consumption in India has grown by about 700% in the last four decades, CORE 2008*)* contributed largely by the changes in the demographic structure brought about through rapid urbanisation, need for socio-economic development, and the need for attaining and sustaining self reliance in different sectors of the economy.

5.1. ENERGY SCENARIO IN INDIA

The decade of the seventies has witnessed major world oil supply disruptions. During the 1970's the OPEC production was cut down by two and a half per cent causing severe oil supply distortions. From 1975 oil prices remained high but not as high as in 1973-74. But the Iranian revolution in 1979 worsened the situation and oil prices again rose sharply in 1979, generating the second oil shock. From the mid 1980s, there was again a resumption of the growth of demand for refined products. This increased demand led to a rise in oil prices from the late 1980s. From July to October 1990, following Iraq's invasion of Kuwait, there was almost a doubling of oil prices. However, this 1990 oil price shock had substantially lesser impact on the world economy than the other two oil price shocks. The reason for this diminished effect was the short duration (only 4 months) of the 1990s oil price hike, inter-fuel substitution such as the substitution of oil, to a large extent, by competing energy sources. In addition, an overall recession of economic activities that had already begun before the price hikes fuelled this effect. India being an oil importing country witnessed significant changes in the energy consumption pattern due to the oil shocks. Inflationary situation arose in India. Faced with rising inflation and a balance of payment crisis in mid 1991 the government of India introduced a fairly comprehensive policy reform package- comprising currency devaluation, deregulation, de-licensing, privatisation of the public sector. The government of India initiated these policy changes to overcome the critical situation. The rising oil import bill has been the focus of serious concerns due to the pressure it has placed on scarce foreign exchange resources(Mukhopadhyay,2002a).

The growth of the Indian economy during 1983-84 to 1991-92 was 5.89%. Indian economic growth has accelerated under the introduction of economic liberalization policies that began in the early 1990s and it continues to show an impressive economic growth till date. India's annual GDP (gross domestic product) growth was about 4% in the early 1990s and rose to 7% roughly in rest of the decade. Recently, country's real GDP grew at an impressive rate of more than 7%. The manufacturing sector has shown a significant improvement in performance. Its averaged rate is about 8.9% during 1990s, which is little higher than the previous decade (it was 6.8% per year during 1981-91). The performance spreads virtually over all aspects of the economy. Together with the country's impressive growth, India has also become a significant consumer of energy resources. In accordance with that primary energy consumption has grown rapidly. To uplift the economy and to accelerate its growth the government of India also initiated new strategies for the energy sector in tune with the economic reforms in mid-nineties. The energy strategies were: i) to initiate a shift from non-renewable sources of energy to renewable sources and to provide adequate energy to the rural and urban poor, ii) to ensure efficiency in the use of energy in all production processes, iii) to review the use of all energy intensive materials and provide for their substitution by less energy intensive materials through Rand D, iv) to ensure efficiency in the use of equipment in the energy sector, especially in thermal and nuclear power generation through improved plant availability, v) to initiate measures aimed at reducing energy intensity in different sectors, through changes in technology and/or processes, vi) to optimise inter-fuel substitution, vii) to propagate renewable resources based on decentralised and environmentally benign non-conventional technologies and viii) to maximise the availability of indigenous energy resources such as oil, natural gas, coal and hydroelectric power, as well as non-conventional energy by way of bio-gas, solar energy and wind energy (GOI, 1992). In view of that here we are providing a brief sketch of the performance of energy demand and supply from 1980 onwards.

According to EIA statistics, India's primary energy consumption grew at an annual rate of 9.4% from 4.04 quadrillion Btu (British thermal units) (10^{15}) in the 1980s to 7.87 quadrillion Btu (10^{15}) in 1990. But the consumption slowed down (16.2 quadrillion Btu in 2005) to 7.04% p.a. from 1990 onwards till 2005 (EIA, 2007). These statistics indicate that India is the world's 5th largest energy consumer after USA, China, Russia and Japan. The energy production in India was high from 1980 onwards (11.99 p.a.) but the trend is declining since 1990s (4.79p.a.). There is a gap always between production and consumption. India is a net energy importer, mostly due to the large imbalance between oil production and consumption. A historical summary of India's total primary energy production (TPEP) and consumption (TPEC) is shown in figure 5.1.

Despite increases in energy use in India, consumption on a per capita basis is still one of the lowest in relation to other countries. In 2000, India's per capita consumption was 12.6 million British thermal units (Btus) against a world average of 65.4 million Btus and a U.S. average of 351 million Btus. Below the chapter is trying to capture the energy situation in India since 1980s.

Coal which accounts for most of the energy supply of India (63 percent of the country's total energy requirements), achieved a steady annual growth of 6.96% during the period.

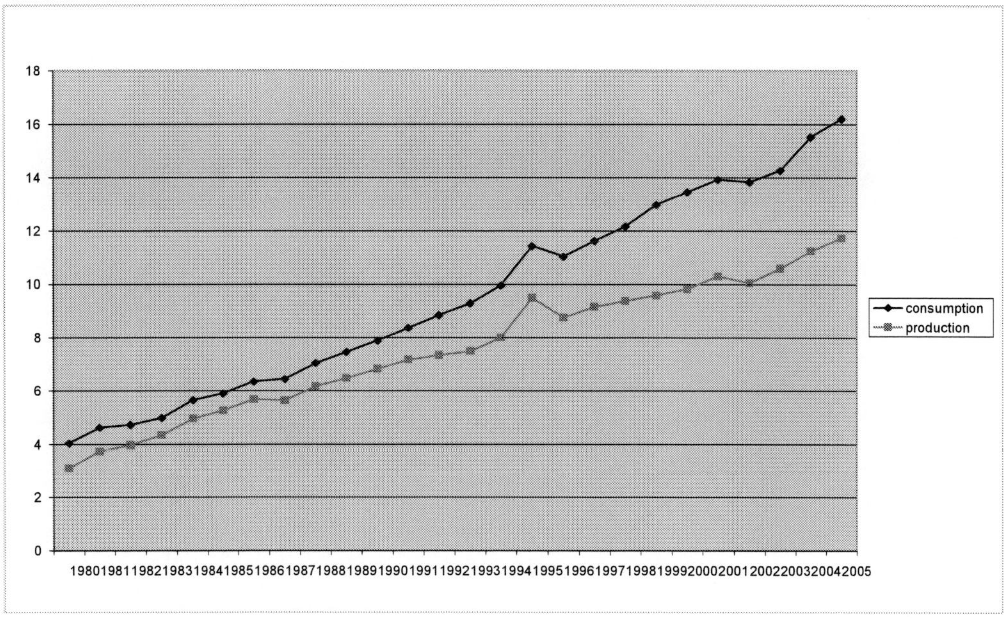

Source: EIA, International Energy Annual, 2007.

Figure 5.1. India's Total Primary Energy Production and Consumption (Btu).

India is one of the largest producers of coal with estimated reserves of 214 billion tons (proven coal /lignite reserves of 118 billion tons). Indian coal has high ash content and needs washing / mixing with imported coal. The power sector alone generates 90 Mt of fly ash as slurry; leaching of heavy metals and soluble salts then leads to groundwater pollution. India is currently the third-largest coal-producing country in the world (behind China and the United States), and accounts for about 8.5% of the world's annual coal production. India is also currently the third-largest coal consuming country (behind the China and the United States), and accounts for nearly 9% of the world's total annual coal consumption. More than half of India's energy needs are met by coal, and about 70% of India's electricity generation is now fueled by coal. The annual demand for coal has been steadily increasing over the past decade, and is now nearly 50% greater than it was a decade ago. Figure 5.2 indicates the trend of production and consumption since 1980s. It shows that the gap between production and consumption becomes widened from 1990 onwards.

Even though India is able to satisfy most of its country's coal demand through domestic production, less than 5% of its reserve is coking coal used by the steel industry. As a result, India's steel industry imports coking coal, mainly from Australia and New Zealand, to meet about 25% of its annual needs (DOE/EIA, 2006). However, coal's share of primary energy consumption has fallen while crude-oil and natural gas gone up, indicating a gradual shift in the Indian energy mix.

Demand for crude oil and natural gas has increased remarkably. The consumption of crude-oil increased at an annual average of 8.1% p.a. during 1980 -90, but slightly declined to 7.05%p.a. from 1990 to 2005 (EIA, 2007). India's oil consumption almost more than doubled from 1168.33 thousand barrel per day in 1990 to 2438.5 thousand barrel per day in 2005 (EIA, 2007).

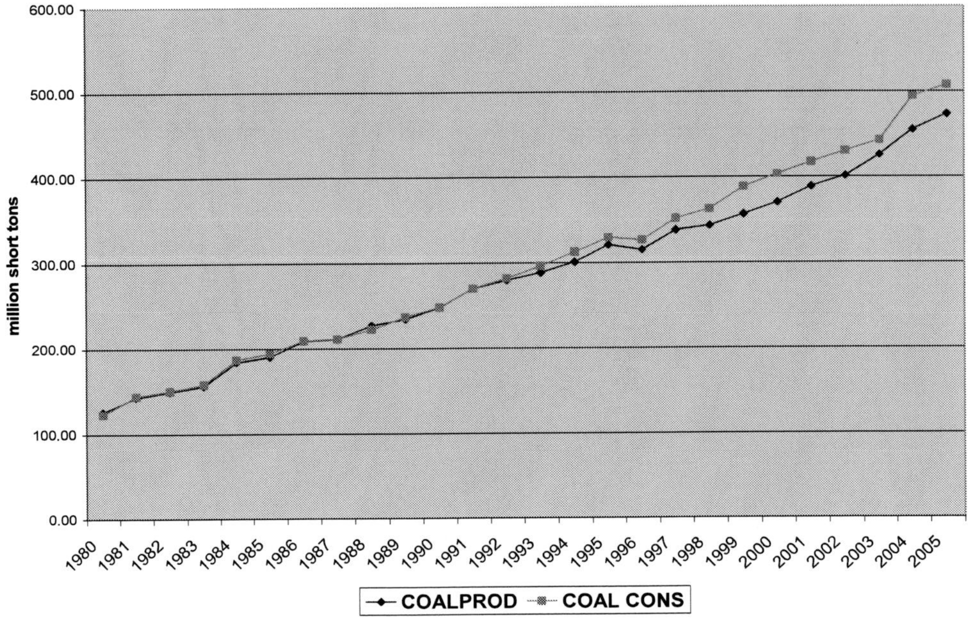

Source: EIA, International Energy Annual, 2007.

Figure 5.2. Coal Production and Consumption in India, 1980-2006.

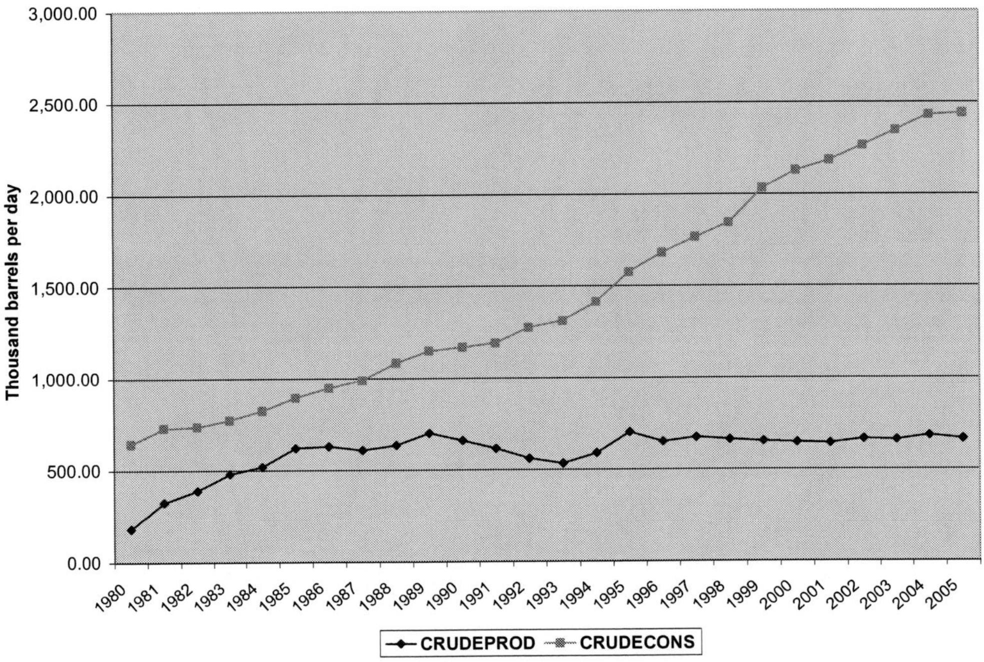

Source: EIA, International Energy Annual, 2007.

Figure 5.3. Crude Oil Production and Consumption in India, 1980-2005.

While India is a fourth largest oil producing nation in Asia following China, Indonesia, Malaysia, its oil output has steadied at around 664 thousands barrel per day in 2005. However, growth of domestic oil production has not been able to catch up with the growth of consumption. India's exploration in the oil sector is inadequate. Nearly 40% of the sedimentary basins still need exploration. The recoverable reserves are estimated as only 640 Mt (a mere decade's supply). According to EIA estimates, India was the fifth largest consumer of oil in the world in 2006. The biggest consumer is the transport sector. So the technologies of fuel cells, energy storage in hydrogen, electricity run vehicles, and replacement of liquid fuels by compressed natural gas are of serious concern to us. The oil imports rose sharply to meet the rapid expanding domestic demand. India's net oil import increased from 508.33 thousand barrel per day(bpd) in 1990 to 1773.4 thousand barrel per day in 2005. During the period, India's dependence on imported oil grew enormously from 43.50% to 70.63%. According to IEA's world energy outlook 2005, India's oil demand is projected to increase by an average annual rate of 2.8% from 2.5 million bpd in 2003 to 5.2 million bpd in 2030. India's natural gas demand will rise 3-5 fold from 28 billion cubic meters in 2003 to 98 billion cubic meters in 2030 (EIA, 2007a). Natural gas now supplies about 7% of India's energy, and has recorded the fastest rate of increase in India's primary energy supply. If continued its share is expected to double by 2020.

Figure 5.4 presents the production and consumption of dry natural gas in India. From the figure it is clear that production and consumption remains same till 2004 but consumption tends to increase after that. The natural gas reserves of India (including a recent big find near the east coast) are 850 billion cubic meters. This is close to 30 year's supply, at current consumption rate. Consumption (especially in power and industrial sectors) is however rising at the rate of 8% per year. Economic exploitation of gas hydrates is a major technical challenge to meet the rising demand.

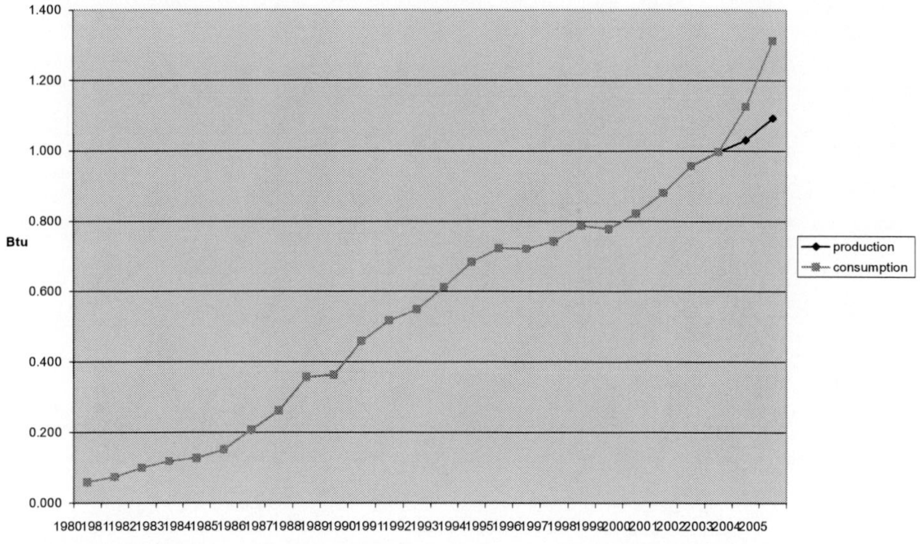

Source: EIA, International Energy Annual, 2007.

Figure 5.4. Dry Natural Gas Production and Consumption in India,1980-2005.

India's need for power is growing at an impressive rate. Electricity generation and consumption in India have both nearly doubled since 1990 (figure 5.5). India is presently the sixth-greatest electricity generating country and accounts for about 4% of the world's total annual electricity generation. India is also currently ranked sixth in annual electricity consumption, accounting for about 3.5% of the world's total annual electricity consumption. Overall, India's need for power is growing at a remarkable rate. Annual generation and consumption have increased by about 64% in the past decade. The projected rate of increase (estimated at 8-10% annually, through the year 2020) for electricity consumption is expected to be one of the highest in the world (CSLF, 2006).

In figure 5.5, Generation consists of conventional thermal electricity, hydroelectric power, nuclear electric power, and geothermal, solar, wind, and wood and waste electric power generation. Consumption includes Net generation + electricity imports-electricity exports-electricity distribution losses.

The brief discussion about the commercial energy shows that the country is having potential in some cases but utilization is not upto the desired level. From the oil front, it is apparent that country has to rely on import. Due to the volatility of the international market country's import bill is rising. On the other hand transmission and distribution losses are making the electricity sector critical. The industrial sector in India is a major energy user, accounting for about 50% of the commercial consumption. There are wide variations in energy consumption among different units within the same industry using comparable technology. The energy saving potential in this sector may be as high as 25% making this sector as having the maximum potential in the economy. Despite the large potential, energy efficiency investments though have financially attractive returns, actually only a small fraction is being tapped.

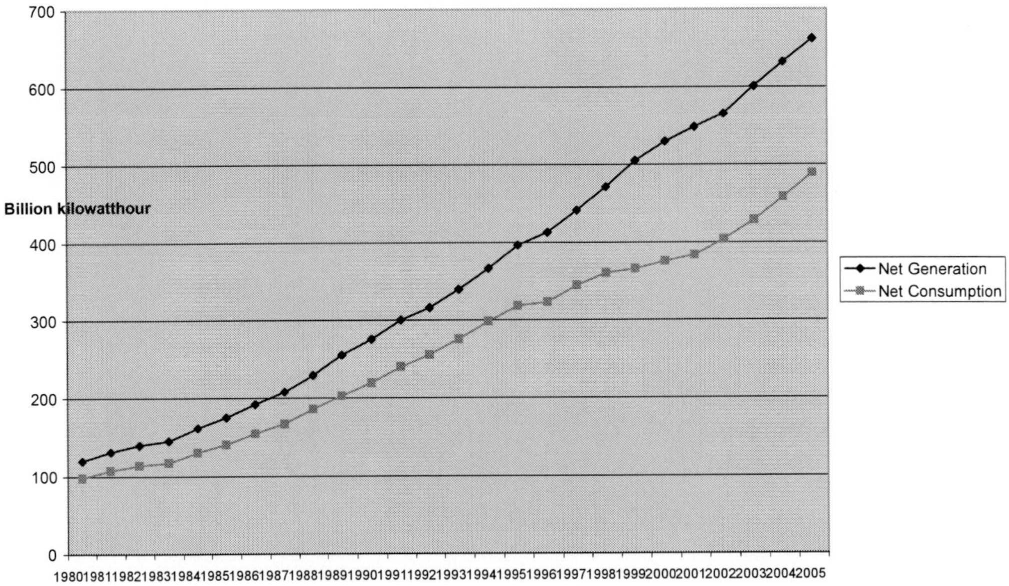

Source: EIA, International Energy Annual, 2007.

Figure 5.5. Electricity Generation and Consumption in India, 1980-2005.

Table 5.1. Percentage Share of Energy Consumption in Different Sectors of the Economy

Years	Energy consumed in agriculture	Energy consumed in industry	Energy consumed in transport	Energy consumed in services
1980-81	9.0	57.0	23.5	13.4
1990-91	4.36	56.01	24.93	14.69
1991-92	4.75	55.89	24.85	14.50
1992-93	4.96	55.69	24.88	14.47
1993-94	5.31	55.43	24.67	14.60
1994-95	5.70	54.74	24.74	14.81
1995-96	5.19	53.53	27.87	13.41
1996-97	5.16	57.14	26.06	11.64
1997-98	5.40	48.01	29.49	17.10
1998-99	4.86	44.04	31.29	19.82
1999-2000	4.79	45.69	32.30	17.31
2000-1	3.53	47.26	32.64	17.56
2001-2	3.02	46.52	33.30	17.75
2002-3	3.33	49.52	31.93	15.27
2003-4	3.85	47.05	32.16	17.27

Source: TEDDY, TERI report various issues.

Table 5.1 reflects the share of energy consumption in different sectors of the economy since 1980s. Transport sector shows high share through out. Though the industrial sector records high share in first half of the 1990s but it starts declining from later half of the 1990s.

Further we know that the energy consumption ---fossil fuel based ---is responsible for environmental pollution. India is currently the fifth-largest carbon emitter in the world (behind only the United States, China, Russia, and Japan) and currently accounts for about 4.2% of the world's total fossil fuel-related carbon emissions (Muneer et al., 2005). There is concern for climate change which has been induced by green house gases owing to use of fossil fuels in generation of energy and transportation. Increasing concern about environmental problems caused by the combustion of fossil fuels has generated a need for knowledge on energy production and consumption patterns.

5.2. Emission Scenario in India

Now we shall present a brief description of the three air pollutants (CO_2, SO_2 and NOx) and their current situation in India.

Among the contribution of different gases to the green house effect, that of CO_2 is the largest. This is not because its potential for global warming is the highest but because of the sheer quantities of CO_2 released into the atmosphere as a result of anthropogenic activities. In the decade of the 1980s, India's carbon emissions have increased by about 12%p.a, and are about nine times higher than they were forty years ago (figure 5.6).

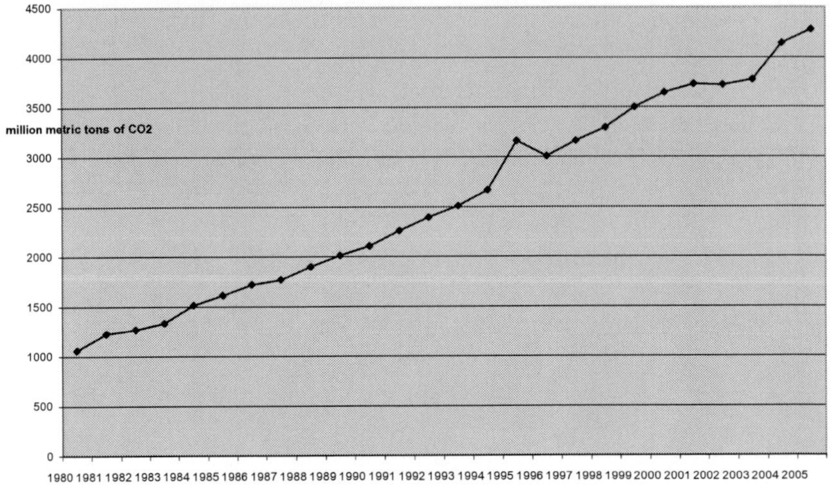

Source: EIA, International Energy Annual, 2007.

Figure 5.6. Carbon Dioxide Emissions in India, 1980-2005.

Between 1950 and 1990, per capita income has increased by 1.6% but per capita emissions of carbon increased by 3.6% annually (TERI, 1997). This increase in emissions reflects primarily an increase in energy use. But in the 1990s and in the current decade the growth of emission slowed to 6%p.a.

Ninety eight percent of CO_2 emissions in India are accounted for by energy related activities. Of this nearly 48% is contributed by bio-mass combustion and the rest by the combustion of fossil fuel. From the combustion of fossil fuels, coal accounts for nearly 70% and oil for 26%, the rest is due to natural gas. The fossil fuel consumption in India is growing continuously, dominated by oil for transport and coal for power production as discussed earlier. Under such circumstances carbon emission can reach an alarming magnitude. Much of this increase is due to India's increasing utilization of its coal resources for power generation. Carbon emissions are forecasted (using a medium economic growth scenario) to grow by about 3.3% annually through 2020. This would cause India to displace Japan as the fourth-greatest carbon emitter by 2010(Hooda and Rawat,2006). India is a non-Annex I country under the United Nations Framework Convention on Climate Change, and as such, is not required to reduce its carbon emissions.

The Indian fuel composition is such that SO_2 and NO_x are also largely derived from the combustion of coal and oil. In case of NO_x, the major sources are coal, gas oil and to a lesser extent, gasoline. The NO_x factor for a given fuel is much higher if the fuel is used in mobile sources rather than in stationary sources. Thus, the NO_x emissions play a dominant role in transportation fuels- diesel driven vehicles are the major source of NO_x contributing to over 90%. Petrol driven vehicles are also the major source of CO emissions contributing to over 85%. Like transport, power sector emissions equally dominate NO_x emissions contributing nearly 30% (Garg et al., 2001). The sectoral composition of SO_2 emissions indicates a predominance of electric power generation sector, residential heating and industrial energy accounting for 70% of SO_2 emissions. In India the annual average concentrations of SO_2 and NO_x are 80 μg/m^3 and 100 μg/m^3 respectively in 1991. Total SO_2 and NO_x emissions were 3542Gg and 2636 Gg, respectively (1990) and 4638Gg and 3462 Gg (1995) growing at annual rate of around 5.5% (Garg et al., 2001). In association with the rising coal use, emissions of SO_2 and NO_x have increased to levels, which pose serious threat to public health and sensitive ecosystems (Chameides et al., 1994).

After presenting an overview of energy and emission in India we shall now devote to the analysis of the empirical results based on different models as developed in chapter 3.

5.3. Results Based on Model 1

Now let us check the emission level of CO_2, SO_2 and NO_x according to our present model. Aggregated hybrid Input –Output tables for the year 1983-84, 1989-90, 1993-94, 1998-99 and 2003-4 have been used to compute direct and total emission coefficient using the equation (**) in model section.

Total Emission

The total industrial CO_2, SO_2 and NO_x emissions (TE) have been calculated using the Input-Output data for the respective years. Total emissions for the three air pollutants were computed as mentioned by the equation (4) in model 1.

Table 5.2. Total Emissions of CO_2, SO_2 and NO_x during 1983-84 to 2003-4(mt of CO_2, SO_2 and NO_x)

Years	Total Emission		
	CO_2	SO_2	NO_x
1983-84	306.883	1.719	5.980
1989-90	474.105	2.728	8.968
1993-94	573.299	3.167	11.339
1998-99	762.626	4.504	13.987
2003-4	982.269	6.077	16.982

The trend of CO_2, SO_2 and NO_x emissions (TE) as shown in table 5.2 can be represented by the bar diagram shown in figure 5.7. It is evident from the diagram that there is an increasing emission trend of all these three pollutants over the decade. The percentage increase for CO_2 and SO_2 emission from 1983-84 to 2003-4 is more than 200%, while that of NO_x is little lower (table 5.3).

Fluctuations in the growth rates of three emissions are observed within this period, however. During the first period the growth was really high for all the three emissions, but in the second period it declined quite a bit in all cases. Again it picked up in the third and fourth period compared to the period 1989-90 to 1993-94.

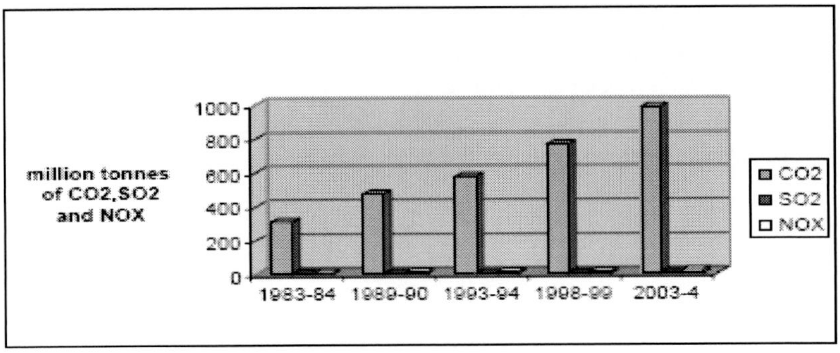

Figure 5.7. Total Emission of CO_2, SO_2 and NO_x during 1983-84 to 2003-4.

Table 5.3. Growth Rate of Emissions in India during 1983–84 to 2003-4

Years	Total Emission		
	CO_2	SO_2	NO_x
1983-84 to 1989-90 (5 years)	54.49	58.67	49.97
1989-90 to 1993-94 (5 years)	20.92	16.09	26.43
1993-94 to 1998-99 (5 years)	33.02	42.23	23.36
1998-99 to 2003-4 (5 years)	28.80	34.90	21.41
1983-84 to 2003-4 (20 years)	220.07	253.47	183.99

5.3.1. Sector-Specific Intensity of CO_2, SO_2 and NO_x during 1983-84 to 2003-4

Now we would concentrate on the sector specific emission intensity for the three major pollutants CO_2, SO_2 and NOx. The calculated sector-specific total intensity of CO_2, SO_2 and NO_x (for 1983-84, 1989-90, 1993-94, 1998-99 and 2003-4) is given in the Appendix D.

Tables (D1-3) show that the overall sector-specific total requirement of CO_2, SO_2 and NO_x is very high for the energy sectors like coal and lignite, crude petroleum & natural gas, and the electricity than that of the non-energy sectors like misc. service, medical and health, misc. metal products and others. Within these three energy sectors electricity has got the highest direct and total intensity of CO_2, SO_2 and NO_x emission across the period.

If we look at the total requirement figure for electricity over time, it shows an increasing trend for CO_2 emission. If we compare first two periods (1983-84 to 1989-90 and 1989-90 to 1993-94) then the contribution leads to higher in the first period and remains same in the second phase. In the third phase (1993-94 to 1998-99) it lowered marginally but in the fourth phase (1998-99 to 2003-4) it records all time high. Electricity sector mainly consumes coal in India. So if the grade of coal is better then it will release less amount of CO_2 and tends to lower CO_2 intensity. Though, India is one of the largest coal producing country in Asia having ample reserves for its future, it is however facing a serious problem of graded coal to be used in the power sector. On the other hand, the demand for electricity is rising to meet the government plan for rural electrification and the "electricity for all in 2020".

The coal and lignite, and crude oil & natural gas intensity has undergone changes overtime. As shown in figure 5.8, there is a sharp fall in coal intensity after 1993-94, and all time lower in 2003-4. The intensity of crude oil & natural gas was reasonable in the pre-reform period, becomes stagnant a bit in the first half of the1990s but turned to be all time high in 2003-4.

The results reveal that electricity has the highest intensity value among all the sectors throughout the period (Appendix D1-4, figure 5.9). For crude oil sector the figure is little higher in 1983-84 but it has declined in the second and third phase and then it starts increasing (1998-99 to 2003-4).

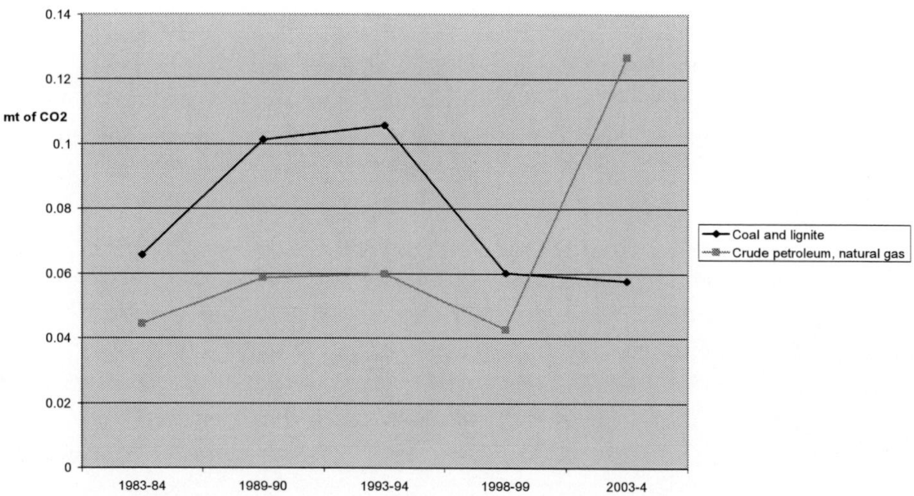

Figure 5.8.Total CO_2 Intensity of Coal, Crude Petroleum & Natural Gas during 1983-84 to 2003-4.

The other prominent non energy sectors having high emission intensity are leather and leather products, rubber products, plastic products, petroleum products, coal tar products, inorganic heavy chemicals, organic heavy chemicals, fertilizers, paints, varnishes and lacquers, other chemicals, gas and water, and other transport equipments.

Some of the important sectors can be discussed in this respect. We observed that the most CO_2 intensive sectors like cement, iron and steel, fertilizer, showing a declining trend while transport sector is fluctuating throughout. The performance regarding carbon intensities in cement sector has improved. It occurs in conjunction with the installation of relatively expensive new technologies such as pre calcining facilities, high efficiency roller mills and

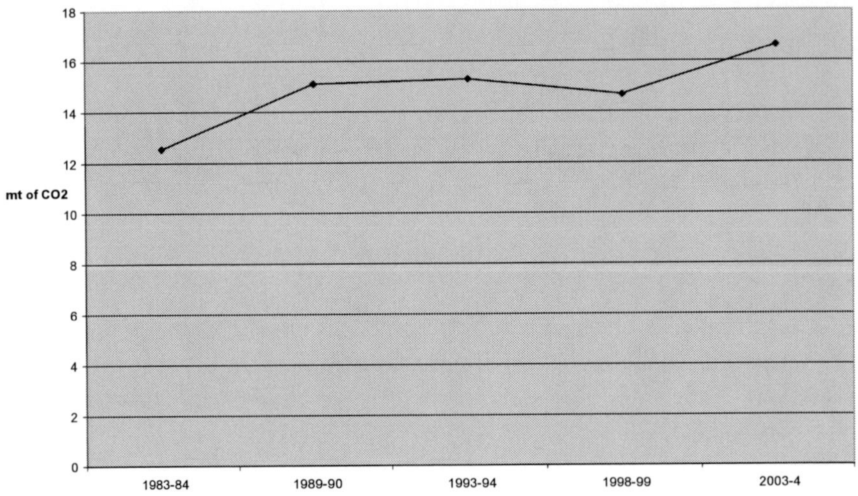

Figure 5.9. Total CO_2 Intensity of Electricity during 1983-84 to 2003-4.

variable speed motor. Actually high efficiency and improved technology lead to low intensity of carbon emission. On the other hand, the increasing trend is observed for construction sector, coal tar products, and petroleum products. Actually construction itself uses many energy intensive materials such as bricks, cement, iron and steel, aluminium glass and asbestos. So the indirect part achieves prominence in this respect leading to high value of total intensity. The above fact carries that the sectors like construction, textile, trade and agriculture, transport emit CO_2 fairly high due to indirect effect. Given the higher value of indirect coefficient and the larger volume of activity, the production of above sectors turn out to be the most responsible for CO_2 emission in India when they are viewed in terms of total (direct and indirect) emissions due to final demand in each sector.

The direct SO_2 emission coefficients were higher generally for the sectors like petroleum products, electricity, chemical and chemical-products, basic metal, metal products, and machinery, trade and other services but these vary between periods. The fertilizer sector is also prominent.

But when we are looking for the total emission coefficients of SO_2 then coal, and crude oil and natural gas should deserve mention. The important sectors are fertilizer, mining and quarrying, iron and steel, cement, and transport, but over the period these sectors show declining trend.

The direct NO_x emission coefficients of electricity record more. Except iron and steel most of the sectors drops a little bit. The total emission coefficients of NO_x emission show very high growth in case of electricity followed by coal & lignite and crude oil compared to other sectors. The sectors like iron and steel, petroleum products, basic metal, machinery transport andcement also record high value.

The emission intensity performance of all the three pollutants can be better viewed from the empirical results of decomposition analysis.

5.3.2. Results based on Structural Decomposition Analysis(SDA)

SDA is a unique technique to study sources of change in the structure of the economy broadly defined by means of a set of comparative static changes in key parameters of an Input-Output table. The total changes in estimated CO_2, SO_2 and NO_x emission from 1983-84 to 2003-4 have been decomposed into effects caused by three components following the equation 11 given in Chapter 3.

Here we consider the comparative static changes of total CO_2, SO_2 and NO_x emissions for the three parameters: *emission intensity(S), technical coefficient (R), and final demand (Y).*

According to the structural decomposition analysis, the change in total CO_2, SO_2 and NO_x emissions between any two years i.e. year o and year t can be considered as 1983-84=o and 1989-90=t in first period, similarly for the second, third and fourth period arrangements are 1989-90=o and 1993-94=t; 1993-94=o and 1998-99=t; and 1998-99=o and 2003-4=t respectively.

SDA for the period 1983-84 to 2003-4

We consider change in total emission due to change in the three above mentioned parameters. The results of these SDA are shown in table 5.4. The total emission change is shown in figure 5.10 and 3rd column of table 5.4.

The total changes of CO_2 and NO_x emissions drop a little during first (1983-84 to1989-90) and second (1989-90 to 1993-94) period. But it increased during later half of the 1990s (1993-94 to 1998-99) and continued till 2003-4 for all the three pollutants. A little bit upward trend has been observed in case of SO_2. The 1st term of the equation (11) i.e., $\Delta S\, R_t\, Y_t$ reflects the CO_2 , SO_2 and NO_x emission changes due to the changes of CO_2 , SO_2 and NO_x intensity of various industries. The values are represented by column (4) of table 5.9. The second term of Equation (11) $S_o\, \Delta RY_t$, defines the CO_2, SO_2 and NO_x emission changes due to changes in technical coefficient matrix shown in the column (5). Similarly, the third term of Equation (11), $S_o\, Ro\, \Delta Y$ refers to CO_2 , SO_2 and NO_x emission changes due to the changes in the final demand of various industries, and the values are represented by column (6).

It is evident from table (5.4) that change in total emission of CO_2 in pre-reform period (during 1983-84 to 1989-90) was mainly due to the change in final demand (around 60%) and then due to change in technological change (around 36%) and the rest was due to the change in intensity level (around 4%). Most importantly, all the factors help to increase emissions.

The scenario drastically changed during the phase, 1989-90 to 1993-94. In reform period, intensity change increases the total emission of CO_2 compared to the first period while change in technological factor reduces total emission. But the effect of change in final demand was consistent and rather more amplified to change the total emission level on the eve of economic reform. What common feature we observed from third and fourth periods is the negative sign of changes in intensity and technology. That reflects that the country is generating less emission per unit of production and using more improved technology in the economy. On the other hand final demand factor increased rapidly thus aggravating the emissions in the economy.

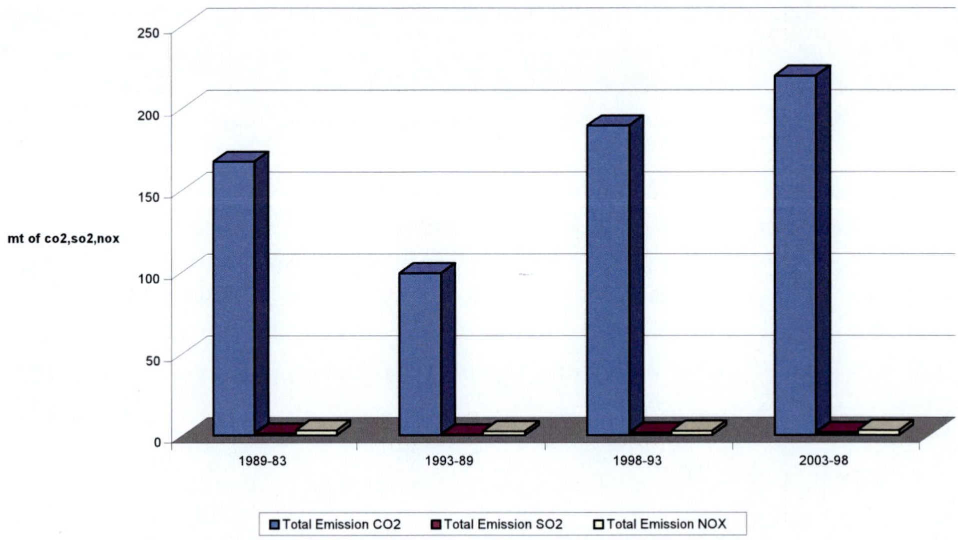

Figure 5.10. Total Emission Changes during 1983-84 to 2003-4.

The favourable contribution of intensity and technology changes helped to reduce emission but the final demand factor on the other hand increased emission more than the reduction. Overall, the final demand rises more than the changes in total emission and also outweighed the contribution of other two factors during 1989-90 to 2003-4. This behaviour of the factors remains same across the three pollutants.

Now we are going to elaborate the factor of CO_2, SO_2 and NOx *intensity* which is shown in table 5.4. As we have seen that changes in intensity throughout the period (1983-84 to 1993-94) became positive for all sectors. It means that the industries are using energy intensive technology or are CO_2 intensive, which in turn contribute to increase CO_2 emission. From direct and total intensity results it is revealed that electricity sector ranks the top among other sectors for CO_2 emission. It is due to the maximum amount low graded coal consumption and also for the inefficient process (Mukhopadhyay and Chakraborty, 1999). The similar performance is also observed in case of transport, iron and steel, and construction sectors. These sectors increase the CO_2 emission significantly. The intensity of SO_2 and NOx emission shows quite a similar picture like CO_2. But the intensity effect turns negative in the next phase. During the period 1993-94 to 1998-99, the economy started changing slowly towards less energy intensive as reflected from the result of (-9.6 mt of CO_2). Similar pattern is also observed for NO_x and SO_2. This trend continued further in the period 1998-99 to 2003-4(-39.14 mt of CO_2). In this period the negative contribution of intensity effect is quite high compared to previous period.

The *rate of technical coefficient* of CO_2, SO_2 and NO_x has been displayed in table 5.4. The changes in the rate of technical coefficient regarding CO_2 emission were positive up to

Table 5.4. Structural Decomposition Analysis of the Emission of CO_2, SO_2 and NO_x during 1983-84 to 2003-4 (mt of CO_2, SO_2, and NO_x)

Year	Pollutants	Total Emission Change (ΔTE)	Comparative Static Change		
			Change in Intensity (ΔS)	Change in Technology (ΔR)	Change in final demand (ΔY)
1983-84 and 1989-90	CO_2	167.222	7.051	59.346	100.825
	SO_2	1.009	0.015	0.299	0.695
	NO_x	2.988	0.277	1.294	1.417
1989-90 and 1993-94	CO_2	99.194	31.893	-107.772	175.073
	SO_2	0.439	0.282	-0.384	0.541
	NO_x	2.371	0.477	-4.855	6.753
1993-94 and 1998-99	CO_2	189.327	-9.581	-14.909	213.602
	SO_2	1.337	-0.250	-1.074	2.690
	NO_x	2.649	0.415	-1.023	3.262
1998-99 and 2003-4	CO_2	219.642	-39.140	-45.780	304.560
	SO_2	1.572	-0.653	-1.097	3.324
	NO_x	2.995	-0.009	-2.438	5.447

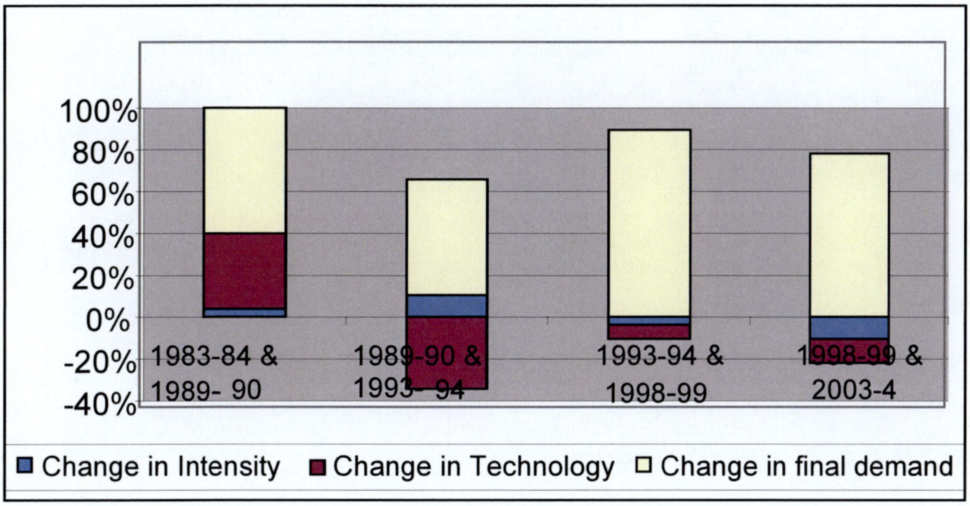

Figure 5.11. SDA for CO_2 Emission during 1983-84 to 2003-4

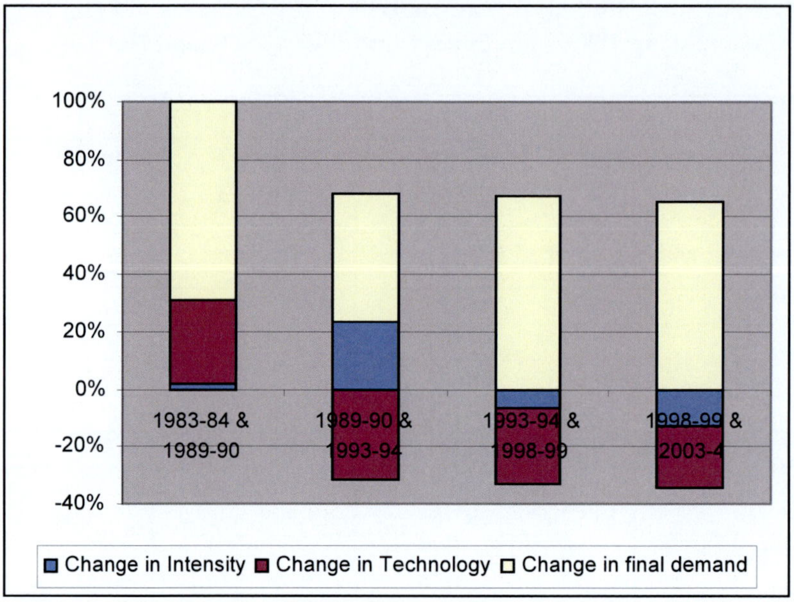

Figure 5. 12. SDA for SO2 Emission during 1983-84 to 2003-4

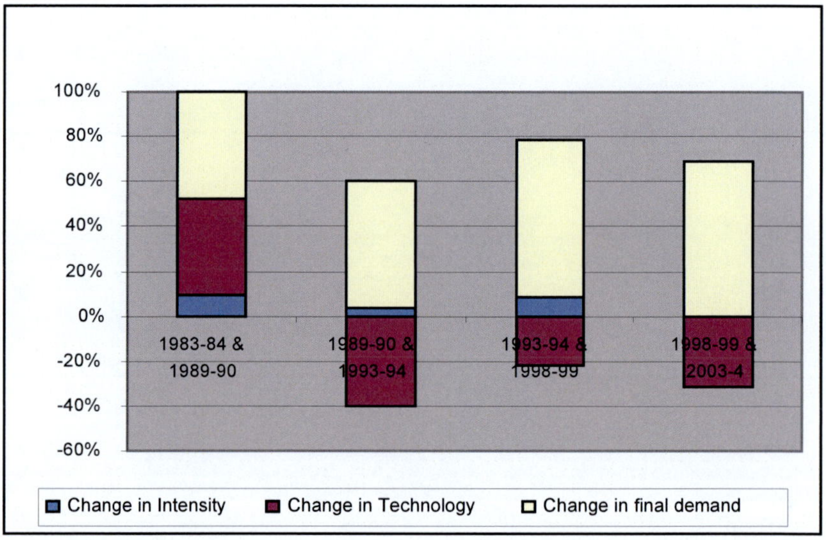

Figure 5.13. SDA for NOx Emission during 1983-84 to 2003-4

1989-90 but in the reform period it became negative. The results of SO_2 and NO_x also reveal the same pattern. This trend continued in the whole study period. The positive or negative influence of any factor on the total emission is better reflected from the performance of the major energy sector coal, crude oil and electricity. The basic reason behind a negative value in the reform period is due to moderate coal and crude oil consumption i.e., 4.8%p.a and 5.6%p.a. respectively during 1991-96 (Mukhopadhyay, 2002). In case of oil sector the technical changes like minimization of the risks of exploration, optimal mix of exploration, energy conservation and inter-fuel substitution have taken place. While in case of coal sector, efficient technology such as exploration, exploitation, efficient utilization, new mining technology have played an important role (Mukhopadhyay, 2002). As a matter of fact we can mention that the technological change has increased emission in the first period. Various reasons can be provided in this respect. Electricity, being a prominent emitter in the economy, can be taken as an example. The low thermal efficiency of power plants in India is caused by the generally small size of its power plants. Besides, the low capacity utilisation of thermal power plants also decreases overall energy efficiency. The average annual load factor of all thermal plants in India was 53.8% in 1990-91. This largely attributed to inefficiency in the operation and maintenance of plants (GOI, 1992).

All these factors worked together to move towards high emission. But the moderate technical changes have taken place resulting in reducing energy consumption which in turn generates low emission in 1990 onwards. The contribution of the fuel sector in this respect goes to coal and oil. New mining technologies for coal have been introduced with a fair degree of success. The slight technical improvement in case of oil and natural gas sector has been made possible due to the flaring of minimization of associated gas, the off take of natural gas, also the minimization of the risks of exploration both by an optimal mix of exploration in different basins in India and vigorous measure for energy conservation and inter fuel substitution. Moreover creation of capacity and its utilisation for oil was very low in 1980s but improved substantially, particularly in the early 1990s. Due to technical improvement in capacity utilisation the growth rate of crude throughput also performed well at 58.6% in 1995-96 which was 4% higher than 1991-92. The alternative technology introduced by the top polluting industries helped to reduce the emission to some extent during the period. Different environmental policies adopted and implemented by the Government of India are in place for such reduction.

The next factor deals with the *changes in the final demand* for CO_2, SO_2 and NO_x emissions. This factor dominates among all other factors. Its contribution was just 75% higher in 1989-90 to 1993-94 compared to previous period (1983-84 to 1989-90) for all pollutants. The contribution slowed down a bit by 21% as compared to the previous and further jumped to 42%. It happened due to high energy consumption by the final demand sector which has increased by 6.9% per year during 1989-90 to 1993-94 (Mukhopadhyay, 2002). The shares of individual sectors are 9% for coal, 5.47% for crude oil and natural gas and 7.85% for electricity in this respect. The demand for electricity in the household sector is expanding rapidly as the pressure of urbanisation continues to increase and the availability of consumer durables also continues to expand. Several of the relatively newer and faster growing industries such as gems and jewellery, garments and electronics are more energy intensive. The rapid pace of urbanization and diverse urban growth pattern involve many basic structural changes in the economy, which have major implication for energy use and also emission. Urbanisation brings changes in the way resources are collected, distributed and used. The rising per capita income associated with urbanization increase demands for both end use energy and energy intensive products and services.

Overall, high contribution made by the changes in final demand factor rather outweighed the contribution made by other two factors- changes in intensity and technology in 1993 onwards. Improved technology and lower intensity started in the economy at a slower pace after the reforms but could not reduce the country's overall emission growth. So the important role played by the final demand factor pushed us further to dig out the matter intensively. Model 2 is formulated towards that direction.

5.4. Results Based on Model 2

In input-output model the final demand vector combines the country's consumption (private and government), investment, changes in stock, export and import. The detail formulation is captured in the model 2 (chapter 3). As we know that the growth of any economy is correlated with the energy consumption, and this has also happened for India. Within the course of the period we have seen that the private consumption in the Indian economy changed drastically. It is biased towards more energy intensive consumption. Since the economy is passing through a phase of transition, and around 30% of the population is living below the poverty line, it would be interesting to see how the different income groups are contributing to energy consumption and emissions. In this section, the study basically tries to look into this matter intensively.

Role of Income Distribution

So far we have not introduced the role of income groups in generation of emissions due to different energy consumption pattern. The sector-wise private consumption expenditure pattern during 1983-84 to 2003-4 is given in Appendix C1-C5. A closer look at sector-wise private final consumption expenditure of different income groups over the study period shows that the higher and middle are walking on the same path. The higher income groups expenditure shows a reasonable increment during 1983-84 to 1989-90. But in the reform period the expenditure boosted up across the different goods. On the other hand, the pattern of lower income groups expenditure across the period does not reveal any drastic change. In the first period, 1983-84 to 1989-90, almost all sectors have shown increase. Though the share of expenditure on few food items and durables like textile has enhanced over the period 1993-94 to 1998-99, but in most of the other sectors the share of expenditure has not increased rather decreased. It is interesting to note that during the period 1998-99 to 2003-4, the share of fuel sector expenditure has increased a bit compared to the previous period 1993-94 to 1998-99, along with minor increase for few durables goods like leather and rubber products, miscellaneous metal products, other transport equipment, construction, and electrical equipment (Appendix C1-C5). This indicates a partial reduction in the poverty level of India[17].

[17] Income poverty has dropped significantly, especially since the early 1980s. Depending on the year, the Gini coefficient of rural consumption expenditure distributions has varied between 0.25 and 0.30, while the Gini coefficient for urban consumption expenditure distributions has varied between 0.32 and 0.34. There has been a decline in the combined poverty ratio from 44% in 1983 to 36.0 percent in 1993 -1994 to 26.1 percent in 1999- 2000. But despite the reduction, the absolute numbers of the poor were quite large, particularly in rural India, where, in 1999- 2000, almost 200 million individuals were BPL (below the poverty line) while 67 million were in urban group. In addition, the rural poverty ratio was higher than the urban poverty ratio. All of this evidence indicates that high growth is a significant force behind poverty reduction. Neit her government poverty alleviation programmes nor other direct policy instruments appear to have made a significant dent in poverty levels on a large enough scale (FAO, 2006). For poverty reduction, it underlines the

In this section we capture the contributions of different factors as described before from the point of view of private consumption made by the lower, middle and higher income groups generating CO_2, SO_2 and NO_x emission.

Changes in Intensity

Changes in intensity made by the three groups are identified in table 5.5. The overall intensity effect sharply increased from first period (1983-84 to 1989-90) to second period (1989-90 to 1993-94). The adjustment of this intensity effect was made by the other final demand sectors like exports, imports, government consumption expenditure. In the first period the other final demand sector acted negatively i.e. the contribution helped to reduce the intensity of carbon, sulfur etc. During the course of the period the effect has changed and moved in opposite directions. This follows that the other final demand sectors are becoming more pollution intensive. It might have been possible due to government final consumption expenditure or export. The increased intensity effect during the study period is distributed among all income groups. Among the income groups, the higher contributes more roughly 60% and the rest is distributed among middle and other final demand sector over the period, while lower income groups are the least contributor throughout. In India approximately 40% of the population belongs to the lower income group, but the contribution towards energy consumption followed by emission is negligible. The performance trend of different income groups to the changes in intensity remains same. The first two periods record positive sign for three income groups which state that consumption of different income groups are more energy intensive. But the situation has changed from 1993-94 onwards. The intensity effect turns negative and helped to reduce energy consumption. The assessment of the whole study period (1983-84 to 2003-4) reflects that the change in the emission intensity of lower income groups fluctuates between 6.31mt of CO_2 to -3.04mt of CO_2 over the period. One interesting feature is the negative role of the other final demand sector throughout. The influence was fairly large in the first two periods but dropped in the last two periods. Overall, it clearly reveals that the reform strategy on energy implemented by the Government of India impacted positively to some extent and is reflected across the different sectors of the economy.

Technological change

We have seen from the results of model 1 that technological change has improved the situation during the period. That is reflected in the contribution of different income groups. A mixed result has been observed from table 5.6. All income groups as well as other final demand vector show a positive sign i.e. it helps to increase emission via the changes in technology in the first period. The trend in second and fourth period is almost similar. The technology effect helps to reduce emission and it is distributed among the different sectors. But the situation is little upset in the third period though the emission of the economy slightly reduced through the technology factor and middle and higher income groups contributed

need for rapid economic development to create more remunerative employment opportunities and to invest in social infrastructure of health and education.

most, the other two contributors responded positively. The middle and higher income groups help to reduce emission due to the technology component throughout, but the role of other final demand and the lower income groups fluctuates in all the period.

Table 5.5. Contributions of Lower, Middle and Higher Income Groups Regarding Changes in Intensity of CO_2, SO_2 and NOx Emissions during 1983-84 to 2003-4(mt of CO_2, SO_2 and NO_x)

Year	Pollutants	Change in Intensity (ΔS)	Comparative Static Change			
			Lower income groups	Middle income groups	Higher income groups	Other final demand sectors
1983-84 to 1989-90	CO_2	7.05	4.77	21.36	39.83	-58.91
	SO_2	0.01	0.17	-1.00	0.77	0.07
	NO_x	0.28	0.09	-0.43	0.63	0.89
1989-90 to 1993-94	CO_2	31.89	6.31	15.42	23.19	-13.04
	SO_2	0.28	0.01	-0.01	0.28	0.01
	NO_x	0.48	0.00	0.38	-0.19	0.29
1993-94 to 1998-99	CO_2	-9.58	-0.82	-3.40	-3.99	-1.36
	SO_2	-0.25	0.00	0.01	-0.13	-0.13
	NO_x	0.41	0.00	-0.19	0.26	0.34
1998-99 to 2003-4	CO2	-39.14	-3.04	-14.54	-19.13	-2.43
	SO2	-0.65	-0.12	-0.20	-0.33	-0.01
	NO_x	-0.01	-0.07	-0.17	-0.22	0.45

Changes in Final Demand

Changes in final demand factor's contribution played a major role among all the responsible factors (table 5.7). Lower income groups' contribution has fallen over the period but middle and higher income groups helped to increase the emission throughout. It gradually increases in each period. The highest is showing in the last period. Compared to middle and lower income groups the major share is grabbed by the higher. The other final demand sector helped to reduce emission in the first period, but for rest of the periods it turns positive.

Table 5.6. Contributions of Lower, Middle and Higher Income Groups Regarding Changes in Technology of CO2, SO2 and NO_x Emissions during 1983-84 to 2003-4 (mt of CO2, SO2 and NO_x)

Year	Pollutants	Change in Technology (ΔR)	Comparative Static Change			
			Lower income groups	Middle income groups	Higher income groups	Other final demand sectors
1983-84 to 1989-90	CO_2	59.35	4.07	1.50	10.01	44.78
	SO_2	0.30	0.07	-0.04	0.64	-0.30
	NO_x	1.29	0.00	-0.06	0.78	0.58
1989-90 to 1993-94	CO_2	-107.77	-9.14	-23.95	-35.43	-39.14
	SO_2	-0.38	0.00	-0.27	0.01	-0.14
	NO_x	-4.85	0.38	-4.45	0.35	-1.17
1993-94 to 1998-99	CO_2	-14.91	-0.88	-10.50	-17.72	14.19
	SO_2	-1.07	0.37	-0.02	-1.03	-0.39
	NO_x	-1.02	-0.53	-1.26	-0.55	1.36
1998-99 to 2003-4	CO_2	-45.78	-2.81	-10.84	-15.79	-16.34
	SO_2	-1.10	0.31	1.20	1.82	-4.43
	NO_x	-2.44	-0.05	-0.45	-1.24	-0.70

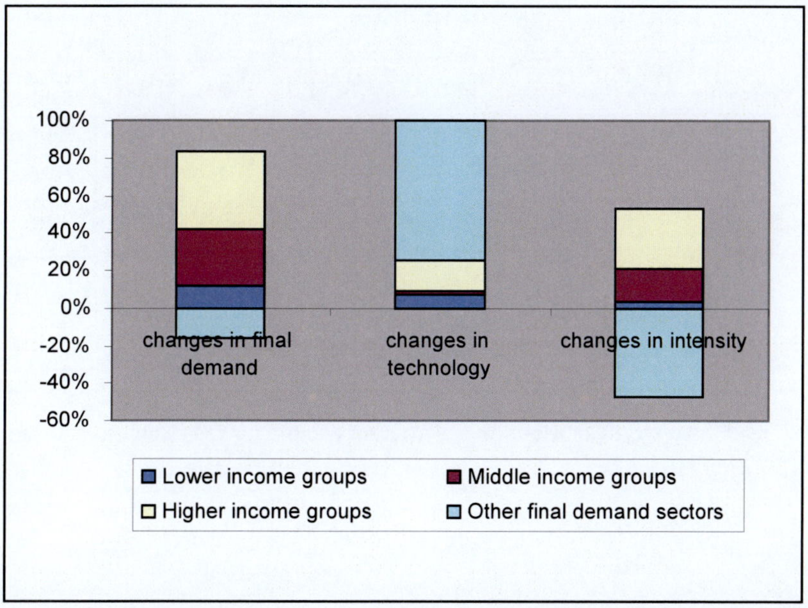

Figure 5.14. Income Groupwise Contributions of Different Factors for CO_2 Emission during1983-84 to 1989-90.

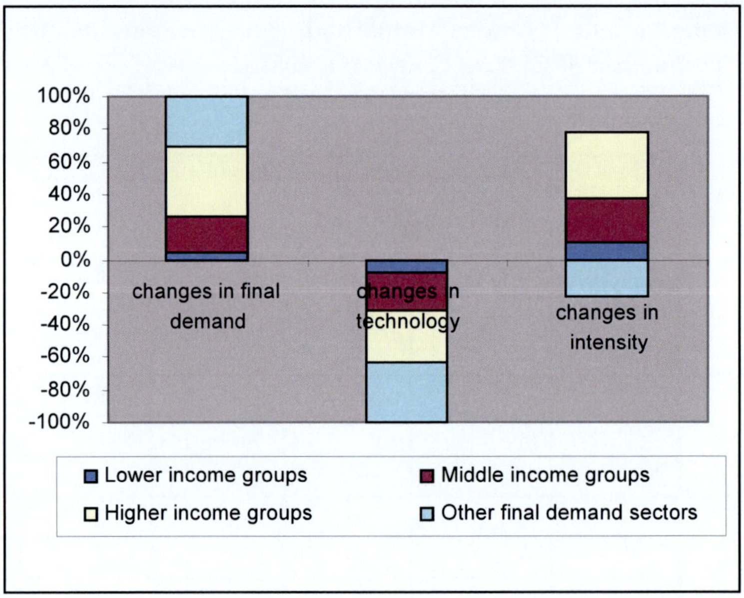

Figure 5.15. Income Groupwise Contributions of Different Factors for CO_2 Emission during 1989-90 to 1993-94.

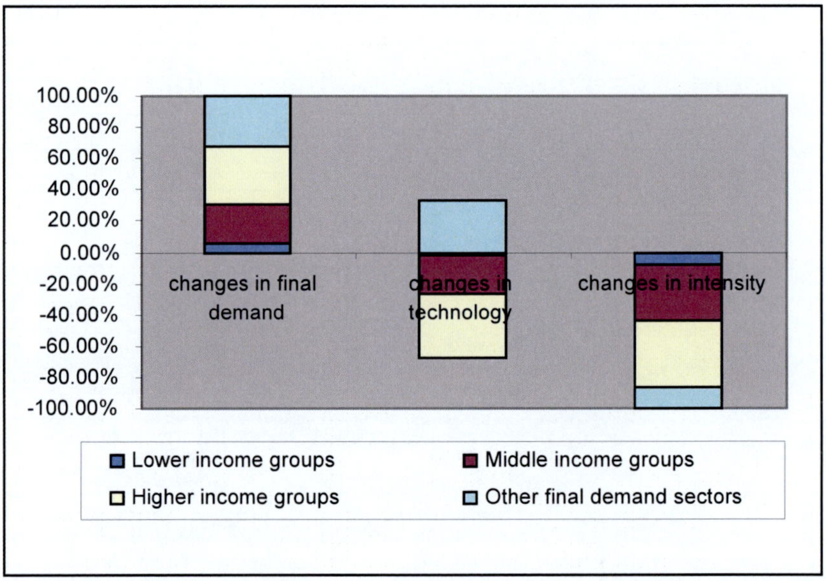

Figure 5.16 Income Group wise Contribution of Different Factors for CO_2 Emission during 1993-94 to 1998-99.

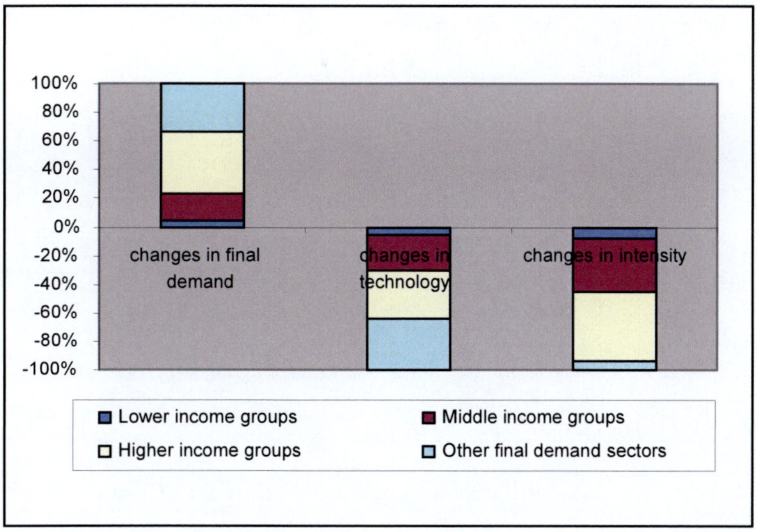

Figure 5.17. Income Group wise Contribution of Different Factors for CO_2 Emission during 1998-99 to 2003-4.

Table 5.7. Contributions of Lower, Middle and Higher Income Groups Regarding Changes in the Final Demand of CO_2, SO_2 and NO_x Emissions during 1983-84 to 2003-4 (mt of CO_2, SO_2 and NO_x)

Year	Pollutants	Change in final demand (ΔY)	Comparative Static Change			
			Lower income groups	Middle income groups	Higher income groups	Other final demand sectors
1983-84 to 1989-90	CO_2	100.825	17.967	44.450	62.631	-24.284
	SO_2	0.695	0.322	0.164	0.454	-0.200
	NO_X	1.417	0.005	0.005	0.589	0.821
1989-90 to 1993-94	CO_2	175.073	8.358	37.705	75.073	53.776
	SO_2	0.541	0.004	0.060	0.277	0.204
	NO_X	6.753	0.178	0.925	2.711	2.943
1993-94 to 1998-99	CO_2	213.601	10.961	53.860	80.894	67.890
	SO_2	2.690	0.012	0.316	1.141	1.216
	NO_X	3.262	0.091	0.804	1.841	0.522
1998-99 to 2003-4	CO_2	304.560	11.830	57.014	132.417	103.299
	SO_2	3.324	0.107	0.329	1.562	1.325
	NO_X	5.447	0.075	0.503	1.499	3.370

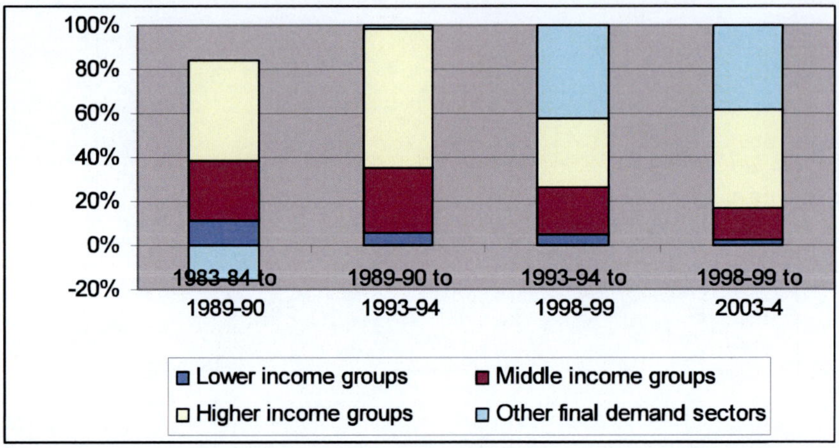

Figure 5.18. Income Group wise Contribution of Combined Factors for CO_2 Emission over the Period

The trend and the pattern of the contribution of different factors remain almost same for all the periods in case of middle and higher income groups. Other final demand component is fluctuating across the factors and periods. The contribution of lower income group is insignificant for all the factors across the period, except the changes in final demand.

Figures 5.14 to 5.17 depict the period wise contribution of each factor in case of CO_2 emission. The degree of shifting between factors for middle and higher is not enormous. Only the striking points to be noted in this respect are intensity and technology effect which help to reduce emission reflected through all the four periods. Emission scenario trend described in the results of model 2 is almost same for CO_2, SO_2 and NO_x.

To get a comprehensive picture of the role of three different income groups across the factors, figures 5.18 is drawn. It captures only the CO_2 performance. Higher income groups dominate for most of the period followed by the middle groups while lower income groups do have a negligible role over the study period. This pattern has also been observed for other two pollutants.

Overall, model 1 shows that the emission of CO_2, SO_2 and NO_x is increasing and change in final demand is identified as a leading factor throughout the study period. As we know that private consumption expenditure plays a major role in final demand factor and it is also reflected from model 2. A closer look at the consumption expenditure table (Appendix C1-C5) shows a marginal improvement of the lower income groups over the period while their contribution in generation of emission is insignificant compared to other two groups. However, the higher income groups' influence is strong throughout on expenditure as well as generation of emission across the factors. The higher income groups are generating more emission (due to high energy consumption) though impacted by the technology and intensity factor favourably. The current study reveals that higher income groups are almost responsible for generating emissions more than 75% and 20-22% is shared by middle income groups. A very negligible amount has been contributed by the lower, however. On the other hand the lower income groups are generally suffering from pollution.

It is well known that the adverse effects of air pollution depend on the level of exposure, the population structure, the nutritional status, and the lifestyle influenced by the income distribution. It is observed (Chapter 1, 2) that the effects are higher in developing nations than developed ones, because the proportion of people living below the poverty line is higher in developing countries. People who live in poverty are those exposed to the worst environmental and health risks. For survival poor people are often compelled to exploit their surrounding, and are exposed to natural resources degradation. Overall, somewhere between 25% and 33% of the global burden of diseases can be attributed to environmental factors (Kjellén, 2001). This proportion is larger in conditions of poverty, where more environmental hazards are present in the nearby living and working environment, and people have less capacity to protect themselves against exposure and effects of harmful and unpleasant pollutants.

Incidence of poverty is high in India and about one third of the population is below the poverty line (belonging to lower income groups) and largely affected by environmental hazards. Now the question arises whether the lower income groups are suffering at all. From what type of diseases do they actually suffer? Are those diseases due to pollution? What will be the share of sufferers due to pollution specific diseases by higher and middle income groups? To address the above queries we have extended our study towards a household survey which has been conducted in selected areas of Calcutta city. The following two chapters are dealing with this.

Chapter 6

A MICRO STUDY BASED ON HOUSEHOLD SURVEY, CALCUTTA

The prime objective of chapters 6 and 7 is to investigate the air pollution effects on human health based on a micro study. In the previous chapter (chapter 5) we analyzed the macro findings based on two models (chapter 3) using data sources described in chapter 4. To corroborate the macro results with micro, the current study collects primary data through household survey in Calcutta megacity. Studies in India cover mainly metro cities like Mumbai, Bangalore, Delhi, and Chennai but a few of them include Calcutta. So the present study focuses on Calcutta.

6.1. ENVIRONMENTAL AND HEALTH RELATED PROBLEMS IN CALCUTTA

Calcutta is one of the growing important metros of India and also in the world, with a total population of nearly five millions. However one of the growing problems with this metro city is the high level of environmental pollution and pollution led diseases are increasing in this city. The Chittaranjan National Cancer Institute (CNCI), one of India's foremost research bodies, reports that diseases include lung cancer, breathing difficulties and asthma. A recent joint study by CNCI, West Bengal's Department of Environment and the Central Pollution Control Board in India have found that around 70% of people in the city of Calcutta suffer from respiratory disorders caused by air pollution. Amongst kids, the figures for respiratory ailments are equally high. The study found that the 56% of the total population's lung function in Calcutta was impaired. The city's highly polluted air is leading to the growing number of lung cancer patients. Calcutta has the highest number of people suffering from lung cancer in the world (WBPCB, 2001). The key finding was a direct link between air pollution among the millions of people of Calcutta and the high incidence of lung cancer. In case of lung cancer in India, Calcutta ranks top - at 18.4 cases per 100,000 people - far ahead of Delhi at 13.34 cases per 100,000 (Bhaumik,2007).

The other air pollution-related health problems including haematological abnormalities, impaired liver function, genetic changes and neurobehavioral problems, were found to be most prevalent amongst those categories of workers exposed to high levels of vehicular emission. These include roadside hawkers, traffic policemen, and taxi and auto drivers. The huge increase in vehicular traffic is one of the basic reasons. WBPCB (2001) shows that epithelial cells also displayed enhanced rate of apoptosis in traffic policemen and garage workers. The results suggest cytotoxic effect of Calcutta's air pollution on nasal chord and epithelial cells in lung. The situation is further compounded in Calcutta with its narrow streets, poorly maintained vehicles, and the use of adulterated fuel. In rural areas, use of bio-fuel causes respiratory problems. The ideal count of Suspended Particulate Matter (SPM) and Respiratory Particulate Matter (RPM) should not exceed 140 and 60 respectively. But Calcutta's average SPM count is 211 and RPM count is 105. Air pollution becomes acute in Calcutta during winter, the ranges are higher than the normal time (900 $\mu g/m^3$ to 1200 $\mu g/m^3$) because the pollutants cannot disperse easily, mainly due to inversion, low wind speed and high congestion. Along with that the worst polluted traffic intersections, this count can be double the city's average during busy hours (Bhaumik, 2007). High SPM in the city air also shows high BSOM (Benzene Soluble Organic Mater). Above all, Calcutta's air pollution results from the horribly high levels of auto emissions which the authorities have failed to control so far.

The West Bengal Pollution Control Board (WBPCB, 2001) reports that twelve Polynuclear Aromatic Hydrocarbon compounds are identified and quantified in the city air and some of them are suspected carcinogen. Among the ten heavy metals, lead concentration in SPM during winter for Calcutta is high in comparison to other cities of the world. The average benzene concentration in winter is 1000 $\mu g/m^3$. Various factors like use of kerosene, coal as cooking fuel, coal in use by power plants surrounding the city, large number of cars, poor quality of fuel, bad condition of the city streets, small road area compared to the total city area, high population density, miserable slum conditions of habitation, and overall poor socio-economic status of city dwellers are responsible for the serious air pollution in the city.

Direct correlation of high particulate concentration in ambient air with higher morbidity and mortality rate among exposed population is well established. A number of studies have been carried out throughout the world[18]. Very few studies have been done in Calcutta. So the present attempt is primarily focusing on Calcutta. The next section explains in detail about the study site.

6.2. Study Site

6.2.1. Description of the Study Site

The City of Calcutta has a total population of 4,580,544 (Census of India, 2001). Calcutta

18 World Bank, 1995 estimates projected 5,700 premature deaths in Calcutta due to air pollution out of 40,000 in the country for the same reasons (WBPCB, 2001). CSE (1997) observes 10,647 premature deaths in Calcutta due to air pollution. In another study Dockery et al., (1994) estimated that due to an increase of 10 $\mu g/m^3$ of RPM in the air, the total mortality rose by 1 percent. A study in 20 US cities by Samet et al., in 2000 calculated that every 10$\mu g/m^3$ rise in RPM increases total mortality by 0.51 percent. A study by Romieu et al.,(1997) in Latin American countries showed that about 1percent increment in total death is associated with every 10$\mu g/m^3$ increase in RPM level. An assessment of health damages from exposure to the high levels of particulates in 126 cities worldwide where the annual mean levels exceed 50 $\mu g/m^3$ reveals that these damages may amount to near 130,000 premature deaths; over 500,000 new cases of chronic bronchitis each year. In aggregate terms, this is equivalent to 2.8 million DALYs lost for this sample of nearly 300 million people or 9 DALYs lost per 1000 exposed residents (Lvovsky,1998).

can broadly be divided into two parts; North Calcutta and South Calcutta. North Calcutta has two major pockets viz; Baranagar and Bidhanagar. From the district census report it has been found that Baranagar has a total population of 250,615 while Bidhanagar has 167,848. South Calcutta has three major pockets which are Gariahat, Jadavpur and Tollygounge. Information from the Kolkata Municipal Corporation suggests that Tollygounge has relatively the highest population of nearly 350,000 habitats. Both the places i.e. Baranagar and Tollygounge have a good mix of people living ranging from high income groups to low income groups. Ambient Air Quality in these two areas is the highest in the city, far exceeding the standard levels. The average ambient air quality in these two areas is given in table 6.1.

Table 6.1. Average Ambient Air Quality in Tollygounge and Baranagar

Particles	SPM (μg/m^3)	RPM(μg/m^3)	SO2(μg/m^3)	NOx(μg/m^3)
Average Emissions	397	248	126	115
Standards	200	100	80	80

Source: http://www.wbpcb.gov.in/html/airqualityinter.php.

From the above facts and figures we see that Baranagar and Tollygounge area are affected with high pollution levels (far above the standards) and population size. Next we investigate how the level of exposure to pollution level is affecting different income classes in Tollygounge and Baranagar area (survey site).

The distribution of the wards under which the survey has been conducted is given in table 6.2. The survey has been carried out in such a fashion that there is a proper mix of households falling under different income classes in different wards.

Table 6.2. Survey of Households in Different Wards of Tollygounge and Baranagar

Tollygounge	Baranagar
92, 95, 97, 98, 120	1, 2, 3, 4, 5, 12, 20, 21, 22, 27, 28

Tollygounge is a typical residential area where the major source of pollution is vehicular emissions. The major road that connects Tollygounge with rest of Calcutta is Netaji Subhas Chandra (N S C) Road. This is particularly very narrow with a continuous movement of the heavy vehicles. Statistics suggest that the average number of vehicles plying on that road per day is 25,000 to 30,000. The people having their residences beside this main road are the worst sufferers. Another problem that the area suffers is the tremendous water logging. Dirty water contamination with drinking water results in various water borne illnesses and skin infections among local residents.

Improper drainage facilities in the localities result in water logging thus making movement very difficult. Rainy season is one of the worst seasons with high water logging problem compounded by slow movement of traffic and thus increased emissions from the vehicles. Estimates of the survey carried out by the West Bengal Pollution Control Board (WBPCB) suggest that the average SPM and SO_2 are over 300 and 100 respectively indicating that the region is highly polluted.

As far as the Baranagar area is concerned it is both a residential and to some extent industrial area. The small industries located are mostly jute mills, few foundries and hardware manufacturing units. The area is connected to the main city by the Barackpore Trunk Road (BT Road). This road is relatively broader than the N S C Road. So the people living in those areas are affected by the industrial as well as the vehicular pollution. Incidentally, the localities are not just beside the main road and thus the effect of vehicular pollution on health is relatively less compared to the Tollygounge area. But this is nullified by the industrial emissions from the firms situated nearby the residential areas.

6.2.2. Survey Methodology

Since number of household is large both in Baranagar (nearly 50,000) and Tollygounge (nearly 35,000), for the survey work we have chosen 90 households in each area spread over different wards to maintain uniformity. The sample is not large (0.18% from Baranagar and 0.13% from Tollygounge). Since we are interested in the different income groups we have considered a sample of ninety households taken from each area viz Baranagar and Tollygounge. The ninety households consist of thirty from high income group (HIG), thirty from middle income group (MIG) and thirty from lower income group (LIG). In consultation with the local councilors and local people -- the residential areas of the HIG, MIG and LIG have been identified. The households were selected on a random basis where the probability that any of the household will be selected has equally likely chance of getting included in the sample. The income group has been made on the basis of the information we got from the survey. It is presented in table 6.3. The LIG households are primarily covered from the slum area in each locality.

Table 6.3. Classification of Groups According to Monthly Income(INR)

SL No.	Classes	Income range
1.	Higher Income Group	> 12,000
2.	Middle Income Group	6,000 – 12,000
3.	Lower income Group	0 - 6000

It is mentionable that the ward number 120 in Tollygounge and 1 in Baranagar are mostly inhabited by the people of lower income groups, although other wards contain a mix of all the income classes. The survey in those two special wards has indicated the general morbid condition in which the poor people live. These areas are given very little importance as far as the health impact is concerned. Apart from air pollution other pollutants in the form of water also affect their life considerably.

6.2.3. General Information Collected

The information gathered from the survey incorporated name, sex, number of members in the family, hours of stay indoors and outdoors, types of houses where they are living, occupational structure, income, expenditure especially on energy like electricity, gas, transportation etc. A good deal of information on health related issues were also covered. Monetary expenditure on health is one of the prime issues that were asked to the households. Apart from this information that plays a vital role in deriving a major conclusion from the survey is the pattern of diseases suffered by the different households. Likewise outdoor pollution, indoor pollution too plays a vital role in determining the disease pattern of the people staying indoors especially the housewives. As a result of this, issue of separate and non separate kitchen has been incorporated. Usage of indoor pollution control devices has also been asked. Emphasis has been given to collect appropriate information of the diseases along with their cause. Moreover, health histories were also collected. This has been done in order to check that there is no confusion between the pollution related diseases and the chronic illnesses. Information on number of health insurances per family was also taken. Some other facts gathered are the number of trips made to doctors and nursing homes and also special cases for hospitalisation and number of work days lost due to illness. The survey was conducted during August to November 2004 based on the detail questionnaire which is attached in Appendix E.

6.3 Features of the Surveyed Household

This section reports the features of the surveyed households of Tollygounge and Baranagar respectively.

6.3.1.Tollygounge

Tollygounge is one of the highly congested areas of Calcutta with a population density of nearly 18000 people per square kilometer. The average family size for the high income group is 3.6, for the middle income group is 3.9 and for the lower income group is 4.5. The total number of male and female interacted under each income category is given in table 6.4. It is evident from the figure that the women interviewed are highest for the middle income group and the lowest from the high income group. Again the number of men interviewed is highest for the high income group and lowest for the middle income group.

Table 6.4. Total Number of Males and Females Interviewed in Tollygounge Area Under Different Income Groups

	HIG	MIG	LIG
Male	20	13	15
Female	10	17	15

Among the males interviewed, it has been found that a lot of the males are chain smokers. Table 6.5 shows that the percentage of smokers in the sample is largest for the LIG (43.47%) and lowest for the HIG (21.73%).

Table 6.5. Percentage of Male Smokers and Non Smokers of Different Income Groups in Tollygounge

	Percentage of Smokers	Percentage of Non Smokers
HIG	21.73	55.00
MIG	34.78	15.00
LIG	43.47	30.00

As far the number of medical insurances is concerned it is highest for the HIG and lowest for the LIG. Figure 6.1 shows the medical insurances of the different income groups.

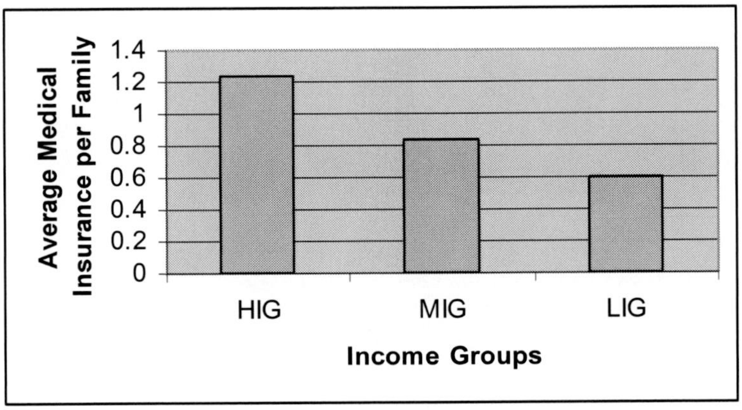

Figure 6.1 Medical Insurances for Different Income Groups in Tollygounge.

Energy Consumption and Health Related Information of Different Income Classes in Tollygounge

Higher Income Group

From the survey analysis it has been observed that the high income groups spend about Rs 1200 per month on energy. The different uses of energy are in the form of electricity, cooking gas, exhaust fan, candles and batteries. Households reside in pukka houses with separate kitchens and proper ventilations in all rooms. The individuals interviewed in this class suffer from various illnesses from headache to shortness of breadth to asthma attacks. Individuals staying indoors suffer mostly from the dry cough and fever and also with skin rashes while the people working outside suffers from headache, shortness of breath and dry scratchy throat. The average number of trips to doctors per month varies from 0 to 5 with an average expenditure of Rs. 500.

In order to prevent themselves from the hazardous effects of air pollution the households in this group adopt various pollution control measures. To prevent outdoor pollution they use gas masks or avoid busy congested roads. They use air conditioners or exhaust fans for keeping indoor air pollution checked. Electric chimneys are also used in kitchens to check the indoor pollution levels. Kitchens are separate in all the surveyed families.

Middle Income Group

The middle income group has a relatively low expenditure on energy consumption. The average expenditure is estimated to be Rs.800. The main source for cooking fuel is observed to be both gas as well as kerosene. 40% of the total households surveyed use kerosene and the rest gas for cooking purposes. Most of the people stay in pucca houses with ventilation facilities. However the kitchens, in 50% cases, are found to be attached with the bed or the drawing rooms and are not built separately. Dry cough, headache and runny nose are some of the diseases which the middle class house wives suffer from. The average number of trips made to a doctor per month ranges from 0 to 4 with an average expenditure of Rs. 450.

Among the individuals who mostly stay outdoors during the day time, headache, cough and chest pain, and respiration problems are quite common diseases. In order to prevent themselves from indoor pollution 40% of the household surveyed have exhaust fans in their kitchens. Most of the interviewed persons working outdoors avoid busy roads. Some of them also use air musk to combat pollution. The average number of restricted days per month ranges from 1 to 4 with an opportunity wage loss of about rupees 300 to 1200. It is mentionable in this regard that the payments to individuals in HIG do not come in the form of no work no pay basis. In the MIG this is to some extent dominant but as we shall see it is most dominant among the lower income groups.

Lower Income Group

This group is the worst sufferer in terms of diseases and exposure to pollution. The average expenditure on energy per month for this group is Rs 600. Kerosene is main source of cooking fuel used along with dung cake and coal. Interestingly, there is no usage of gas, and candles are used for lighting purposes with very little or no usage of electricity. There is no proper ventilation in the houses. The houses are kucha and the kitchens are not separate. Most of them stay in a single room house. So the level of indoor pollution is expected to be the highest. The people staying indoors and outdoors are more or less exposed to same level of pollution, so the disease patterns are also same. The common diseases that the people suffer from are headache, throat irritation, fever, respiratory problem, dry cough, chest pain, and heart disease. To prevent themselves from indoor pollution the household cannot afford expensive pollution control devices as we have seen for the high income classes. The number of trips made to local doctors or physicians per month ranges from 0 to 5. As a result the average monthly expenditure is Rs.430 which is very close to the expenditures of the household belonging to HIG and MIG. The average people in the lower income groups cannot afford their medical expenditure. Some of them are able to borrow money from workplace, but most of them failed to get. In this circumstance, it is difficult for them to sustain.

Comparison between the Different Income Classes with Reference To Several Indicators In Tollygounge

In this section we are now making a comparative analysis of the different income groups with regard to several indicators on the basis of the information provided in the previous discussion. From the field survey it is clear that the percentage share of expenditure on health by low income groups is highest while lowest for high income groups. This is evident from table 6.6

Table 6.6. Share of Expenditure on Health for Different Income groups in Tollygounge

Income Groups	Percentage Share
HIG	3.14
MIG	5.04
LIG	15.52

We can see that the expenditure share for the low income class is 15.52% while that for the high income class is 3.14%. Again if we measure the number of work days lost due to health related problems the results are also in the same tune. The number of work days lost is highest for the LIG and the lowest for the HIG (table 6.7).

Table 6.7. Work Days Lost for Different Income Groups in Tollygounge

Income Groups	Average Work Days Lost Per Month
HIG	2.89
MIG	4.14
LIG	6.08

As far as the expenditure share on energy is concerned table 6.8 reports that the high income class has the highest share while low income class having the lowest.

Now as far the sufferance of the people from pollution is concerned it is the individuals from the LIGs who are suffering the most. This is evident from figure 6.2.

We observe that the percentage of people suffering from respiratory disease is highest for the low income group and lowest for the high income group.

Table 6.8. Share of Expenditure on Energy for Different Income Groups in Tollygounge

Income Groups	Percentage Share
HIG	10.73
MIG	9.40
LIG	7.97

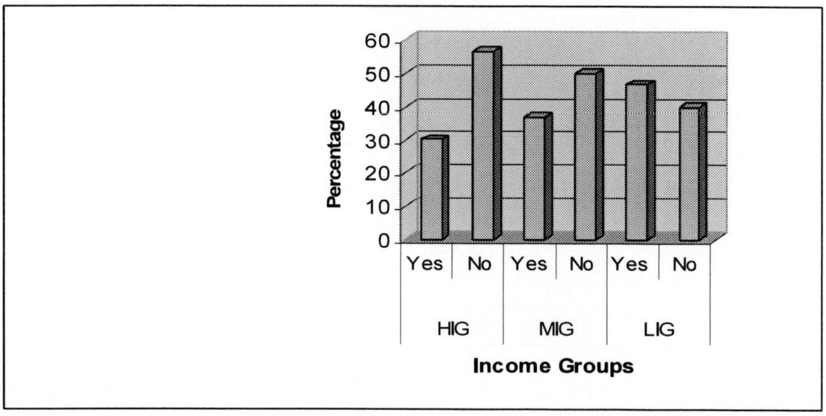

Figure 6.2. Sufferance of Household from Respiratory Disease for Different Income Groups in Tollygounge.

We also need to see the disease pattern for both male and female for different income groups. It has been observed that the housewives are the worst sufferer from indoor pollution and the working members of the families (in most cases it is men) are the worst victims from the outdoor pollution. Figure 6.3 shows how the male and the female are affected from the pollution for the different income groups.

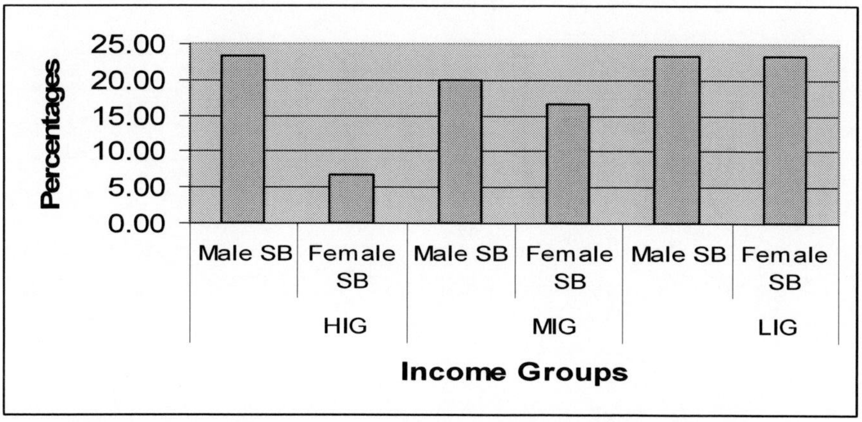

Figure 6.3. Sufferance from Shortness of Breath (SB) of Different Sex in Tollygounge.

It is also evident that the percentage of women suffering from shortness of breath is highest for the LIG. Similar is the case for men. The percentage of female population suffering from the same disease is the least for HIG. As far as the men are concerned it is least for the MIG.

6.3.2 Baranagar

Let us now take a look at Baranagar which is also highly congested area of Calcutta with a population density of nearly 15,000 people per square kilometer. The average family size for the high income group is 4.2, for the middle income group is 4.4 and for the low income group is 4.6. The total number of male and female interacted under each category for the different income groups is represented in table 6.9.

Table 6.9. Total Number of Males and Females Interviewed in Baranagar Area Under Different Income Groups

	HIG	MIG	LIG
Males	14	11	16
Females	16	19	14

The number of females interviewed is highest for the MIG and lowest for the LIG. On the other hand the number of males interviewed is highest for the LIG and least for MIG. Like Tollygounge, the survey found no female smokers in Baranagar also. Survey suggests that the number of smokers is highest for LIG and the least for MIG. This might be because of the health consciousness among the richer class of the society. Table 6.10 shows that the percentage of smokers in the sample is largest for the LIG and MIG and lowest for the HIG.

Table 6.10. Percentage of Male Smokers and Non Smokers of Different Income Groups in Baranagar

	Percentage of Smokers	Percentage of Non Smokers
HIG	20.00	39.21
MIG	40.00	29.41
LIG	40.00	31.37

As far the number of medical insurances is concerned it is highest for the HIG and lowest for the LIG. Figure 6.4 shows the medical insurances of the different income groups.

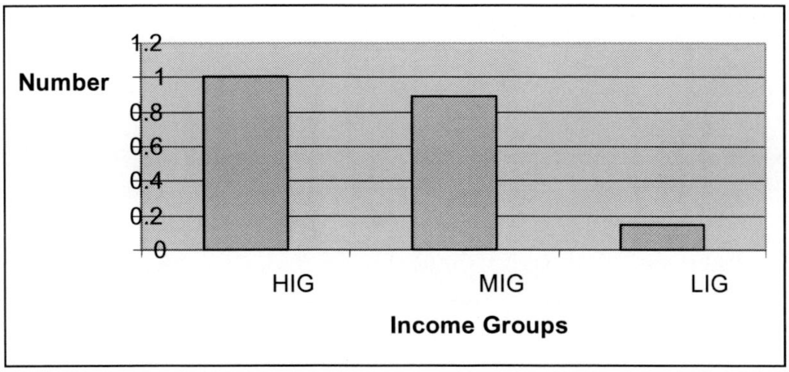

Figure 6.4. Medical Insurances for Different Income Groups in Baranagar.

The average number of medical insurance for the high income group is one. Again for the middle income class the value is 0.89 and finally for the low income class the value is 0.14. This pattern was also observed for Tollygounge.

ENERGY CONSUMPTION AND HEALTH RELATED INFORMATION OF DIFFERENT INCOME CLASSES IN BARANAGAR

Higher Income Group

Like Tollygounge, Baranagar's high income groups also use different types of energy in their day to day life in the form of electricity, cooking gas, batteries, candles etc. The average monthly expenditure on energy for this group is Rs.1100 which is slightly less than what we have seen in Tollygounge. The individuals reside in pucca houses with separate kitchens. Moreover they use various devices to check the indoor pollution from cooking. The devices are exhaust fan, electric chimneys, greater number of windows in kitchen. There is also proper ventilation in other rooms. Among the people who are staying indoors especially the housewives suffer from cough, headache and runny nose. The office goers, not only suffer from headache, but also from burning throat and eyes which results in fevers and influenza, shortness of breath and asthma. The average number of visits to the doctors per month ranges from zero to four. The medical expenses are nearly Rs. 600 per month. The medical expenses range from as low as Rs. 40 to Rs. 2000 monthly.

In order to protect themselves from the hazardous affects of air pollution (both indoors and outdoors) the people in this group take various pollution control measures. In order to face minimal outdoor pollution people use gas masks or avoid busy congested roads. The persons using own cars have air conditioners fitted in their cars with dust suction machines. Some of the families are also seen to use oxygen purifiers in their houses along with room air conditioners.

Middle Income Group

The middle income group has monthly energy expenditure of Rs. 800. The families are observed to use both LPG gas and kerosene for cooking. Electricity, candles are used for lighting purposes. In few cases kerosene used lamps are also used. Most of the houses have separate kitchens. The families mostly live in pukka houses. However, all the houses do not have proper indoor pollution controlling devices generating from the kitchens. Among the devices used, exhaust fan is mostly used one. Out of the total households surveyed only one household was observed to use electric chimneys. The number of windows for ventilations in kitchen is also less compared to the high income group. The office goers usually avoid busy congested roads. Some of them also use masks for avoiding heavy intake of toxic gases.

Individuals in this class suffer from various illnesses such as headache, shortness of breadth, asthma attacks. Individuals staying indoors suffer mostly from the dry cough and fever. One very common feature among the housewives is the excessive sneezing during cooking activities thus indicating that less attention is given to cooking related pollution. The housewives also suffer from skin rashes while the people working outside suffer from headache, shortness of breath, burning eyes, and throat, and flue. The number of trips made to doctors ranges from zero to six. The individual families on an average spend rupees four hundred and fifty per month for medical expenses.

Lower Income Group

This group has the least expenditure on energy. Estimates suggest that the average monthly expenditure on energy is Rs.500. Most of the families use kerosene for cooking purposes although one or two families use cooking gas. Candles and kerosene lamps are mostly used for lighting purposes. However, some of the families also use electricity for the same. The relative low expenditure on energy can be attributed to the little or no use of cooking gas and also little usage of electricity. The houses are a mix of kucha and pucca types. The kitchens are not separate and do not have proper ventilation. There are no exhaust fans. In most of the cases bed and kitchens are in common place. In some cases, the bedrooms are close to kitchens which results in the complete spreading of the pollution related to cooking activities. Here also we see that the people staying indoors and outdoors are more or less exposed to same level of pollution so the disease patterns are also same. The common disease that the people suffer from is headache, throat irritation, burning eyes, fever, respiratory problem, dry cough and chest pain. To prevent themselves from indoor pollution households neither afford expensive pollution control devices as we have seen for the high income classes nor separate rooms. Housewives who are engaged in cooking activities suffer from burning eyes and acute sneezing indicating the inadequate arrangements for reducing indoor pollution. The average number of trips made to a doctor per month ranges from zero to four with an average expenditure of rupees three hundred.

For controlling outdoor pollution the only activity that the people in this group can undertake is to avoid busy areas. They do not use any masks. Neither can they take the pleasure of staying indoors to prevent themselves from the exposure of toxic gases.

COMPARISON BETWEEN THE DIFFERENT INCOME CLASSES WITH REFERENCE TO SEVERAL INDICATORS IN BARANAGAR

For Baranagar the results of the survey are also in tune with those of Tollygunge. The households in the lower income groups are also observed to have the highest expenditure share on health compared to that of the middle income group and the high income group.

The diagrammatic representation is given in figure 6.5

The expenditure share for the LIG is more than eight percent. As far as the middle income and high income groups are concerned the shares are less than four percent.

The number of work days lost on an average per month also shows similar patterns. The LIG has the highest number of work days lost due to illness while HIG has the lowest. The average work days lost for the HIG are 3.29, for the MIG and LIG the figures are 4.44 and 6.88 respectively.

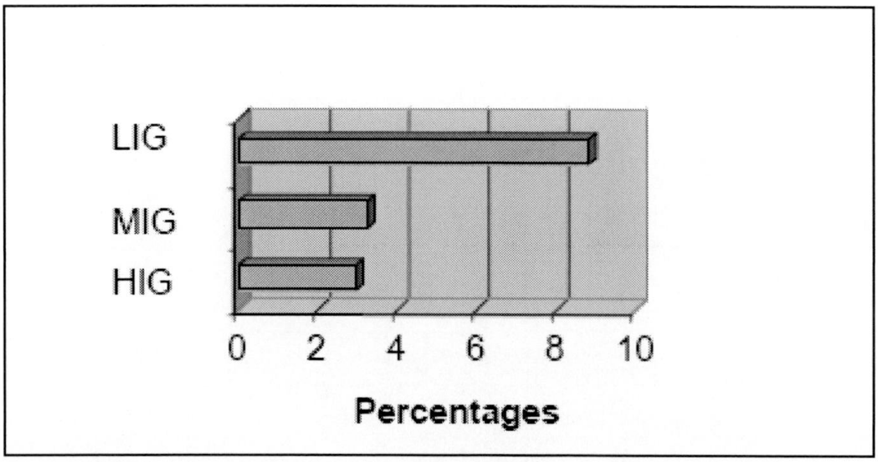

Figure 6.5. Share of Expenditure on Health for Different Income Groups in Baranagar.

Table 6.13. shows that the share of expenditure on energy is highest for the HIG (7.68%)

Households from the lower income group have the highest incidence of respiratory diseases. About 65% of the total households surveyed responded positively as the sufferance from shortness breath is concerned. In the middle income group less than 40% of the total surveyed households informed that they are suffering from respiratory diseases. The value for the high income group is lowest having less than 30% suffering from the disease while it is around 50% for LIG.

Breaking up the disease burden into different sex for each income group, the survey shows that it is the males who are suffering more than the females for each income group. For LIG the percentage values for the male and female are 36% and 25% respectively. The male and female percentages for the MIG are 30%, 20% while for HIG it is 20%, 17% respectively.

Table 6.12. Work Days Lost by Different Income Groups in Baranagar

Income Groups	Average Monthly Work Days Lost
HIG	3.29
MIG	4.44
LIG	6.88

Table 6.13. Share of Expenditure on Energy for Different Income Groups in Baranagar

Income Groups	Percentage share
HIG	7.68
MIG	6.74
LIG	5.9

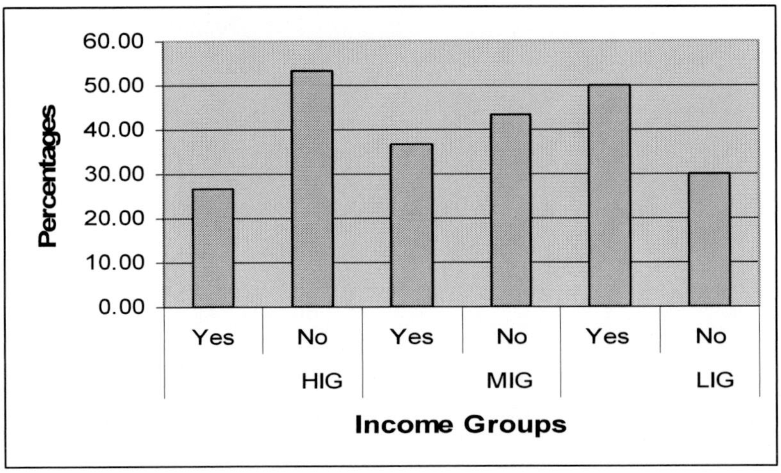

Figure 6.6. Sufferance of Household from Respiratory Disease for Different Income Groups in Baranagar.

Thus the above findings of the survey depict the major features of the household of different income groups in Tollygounge and Baranagar. The survey data also indicate the relationship between air pollution generated from the energy consumption and its related impact on health. The next task is to establish the relationship using the survey data. For the proper economic analysis, dose response model is applied for the study. The study utilized the findings from the survey to calculate how the pollution is affecting the general health conditions of the surveyed people in the following chapter.

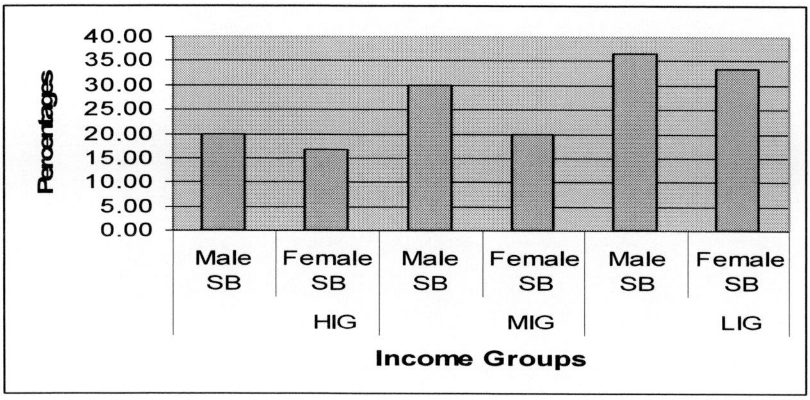

Figure 6.7. Sufferance from Shortness of Breath (SB) of Different Sex in Baranagar.

Chapter 7

MODELING THE HEALTH EFFECTS OF AIR POLLUTION

It has already been seen in chapter 5 an increasing trend of CO_2, SO_2 and NO_x emissions in India. Within 20 years, the emissions increased more than 200% for CO_2 and SO_2 and 183% for NO_x. In chapter 1, the study described about the different pollutants and the likely impacts on health, i.e., the type of diseases due to the pollutant exposure and also explained the diseases due to SO_2, NO_x and CO exposure. Actually carbon monoxide is produced by the inefficient combustion of carbon fuels. Our result follows that so far as the CO_2 is gaining its importance by amount then definitely the inefficient combustion will also be higher and the amount of CO will also get rise. Thus ultimately CO_2 in form of CO can pose a serious health problem in India. Moreover, the level of CO in different parts of India shows a rising trend (chapter1).

In this chapter we have utilised the surveyed data (detail in chapter 6) to estimate how the health condition of the surveyed households is affected due to air pollution in a metro city of India, Calcutta. Moreover, a detailed picture of the surveyed area in chapter 6 shows that how acute is the problem of pollution as well as its impact on health.

We have adopted the dose response function by applying logit model to see the relation between pollution and health. The households surveyed have been divided into three major income groups-- higher, middle and lower. Using relevant information from the questionnaire we have developed two dose- response models. They are (i) Dyspnea or shortness of breath and expenditure on energy consumption for different income groups and (ii) Dyspnea or shortness of breath and direct Exposure (DE) to pollution. We have used simple logit model for deriving the simple relationship between pollution, energy expenditure and general health conditions. Before going to the estimation and discussion of the results, we reviewed few literatures especially that dealt with dose response function.

7.1. LITERATURE ON DOSE RESPONSE STUDY

Research in the field of environmental science and economics has seen a number of studies on the economic valuation on health impact of air pollution. One of the commonly used statistical methods is dose response function in health literature. It is mentionable that logit analysis is a good measure of developing the dose response function. Here we shall briefly review some of the studies with dose response method.

Walter and Fellner (1978), Mendelsohn and Orcutt (1979), Fadel and Massoud (2000) should deserve mention in this respect. Walter and Fellner (1978) have used exploratory techniques for the determination of potential dose-response relationships between human health and air pollution. Mendelsohn and Orcutt (1979) have made an empirical analysis of air pollution dose-response curves. Mullahy and Portney (1990) used data set to explore the links between urban air pollution, particulate matter and chronic respiratory illness on the other. The results indicate that ambient ozone concentrations may be associated with sinusitis and hay fever, while particulate matter may be associated with more serious respiratory diseases. The logit analysis has been used in the study.

Krupnick (1997) explored the appropriateness of concentration-response function transfers by comparing two health studies in Los Angeles and Taiwan. Daily records from a diary-type epidemiological study are used to fit logit equations predicting the probability of experiencing minor acute respiratory symptoms as a function of pollution and weather variables, individual characteristics, and health background and proxies for reporting effects. Zuidema and Nentjes (1997) estimated the dose–response relationship between air pollution and the number of work loss days for the Netherlands. The study is based on illness data (work loss days) for the Dutch labour population and average year concentrations of air pollution in 29 districts. The dose–response relationship has been estimated by means of two different techniques: the ordinary least squares method (OLS) and the one-way fixed-effects method (OWFEM). In general health effects are much smaller when OWFEM is applied than if OLS is used. With OWFEM a significant relationship is found between sulphate aerosol (SO_4), ammonia (NH_3) and the number of work loss days (WLDs). Particulates (TSP), O_3 and SO_2 have no significant effect on the number of WLDs. These results differ from those obtained in studies in the United States, which indicate particulates (TSP) and other small particles, ozone (O_3) and to a lesser extent SO_4 and SO_2 significantly influence the number of WLDs.

Eskeland et al. (1998) estimate dose-response functions for respiratory disease among children based on data from public clinics in Santiago. They find that respiratory disease among Santiago's children is significantly affected by air pollution, measured as PM_{10} (small dust particles). Inclusion of a wide range of covariates and alternative specifications does not change the conclusion. In some model specifications, ozone is also found to affect respiratory illness.

Hansen et al. (2000) investigated the effects of the relationship between air pollution and human health effects. They studied the reduced labour productivity in terms of sick-leaves, which is an important factor in assessment of air pollution costs in urban areas. For this purpose they employed a logit model along with data on sick-leaves from a large office in Oslo and different air pollutants. The results indicate that sick-leaves are significantly associated with particulate matter PM, while the associations with SO_2 and NO_2 are more ambiguous. They also estimated the induced social costs in terms of lost labour productivity and increased governmental expenditures, although these estimates are more uncertain.

Murray et al. (2001) tried to answer whether the dose-response models that are based on research conducted in developed countries can be applied to exposures to air pollution in developing countries. The study considered this issue and examined the factors that may lead to either increased sensitivity or increased human tolerance of air pollutants. It is suggested that although there are factors in developing countries that may increase or decrease human sensitivity to air pollution, overall, a similar range of sensitivity can be expected by individuals in these countries responding to the same effective dose as those in developed countries. As we know that major works have been conducted on developed countries due to the availability of reliable data. But in developing countries also the method has been applied.

A considerable amount of work has also been done in the areas of air quality index and use of dose response function in India. Lioy (1990) has assessed the total human exposure to contaminants using suitable air quality index in different polluted environments. This study has shown how the exposure to different level of pollution affects individual receptivity to diseases. Srivastava and Kumar (2002) have used air quality index to calculate the direct exposure to different level of pollution and developed a dose response relationship to calculate the change in the probability of disease occurrence with a unit change in the exposure. They have also made an economic valuation of morbidity and mortality through lost salary approach. The results show that the avoidance cost is 29% of the total health damage cost. Parikh (2001) has also carried a study on data based on Chembur hospital, Mumbai to calculate the same. Moreover she has made an economic valuation of the morbidity and mortality for the lower income group of the same area. Pandey et al. (2005) estimated human risk (HR) for three pollutants, namely, suspended particulate matter (SPM), nitrogen dioxide (NO_2) and sulfur dioxide (SO_2) for Delhi City. For estimation and analyses, three zones have been considered, namely, residential, industrial and commercial. The total population has been divided into three age classes (infants, children and adults) with different body weights and breathing rates. The exercise takes into account age-specific breathing rates, body weights for different age categories and occupancy factors for different zones. Results indicate that health risks due to air pollution in Delhi are highest for children. For all age categories, health risks due to SO_2 (HR_SO_2) are the lowest. Hence, HR_SO_2 has been taken as the reference with respect to which HR values due to SPM and NO_2 have been compared. Taking into account all the age categories and their occupancy in different zones, average HR values for NO_2 and SPM turn out to be respectively 22.11 and 16.13 times more than that for SO_2. However work on Calcutta in this field is rare. The current study investigates the level of exposure of air pollution affecting the health of the different income classes in selected areas of Calcutta.

7.2. Econometric Specification And Estimation

The logit model estimates the probability of a certain event occurring, given a certain set of independent variables. The dependent variable (y^*) is a latent variable and is not observable. What we observe is a qualitative variable y_i, defined by

$$y_i = 1 \text{ if } y^* > 0 \qquad (1)$$

$$= 0 \text{ otherwise}$$

Let Pi be the probability of $y_i = 1$, i.e., in our case, the subject is suffering from shortness of breath.

The error term is assumed to be logistic and we obtain

$$\text{Log}\,(Pi / 1\text{-}Pi) = \beta_o + \text{Sum}\ \beta_i{*}X_i \text{ for } i = 1, 2....K \quad (2)$$

$$\{\delta\, Pi / \delta\, Xi\} = \beta_i * Pi{*}(1\text{-}Pi) \quad (3)$$

The left-hand side of the equation (2) is called a log-odds ratio. The log-odds ratio is a linear function of the explanatory variables. After estimating the parameter, we can predict the effect of changes of any of the explanatory variables on the probability of the observation belonging to either of the two groups e.g., equation (3).

In the dose-response function we have two options for two choices (having a symptom or not) to analyse the effect of change in air pollution levels on the population having shortness of breath so that a simple logit formulation can be employed. We form the logit analysis for the three income classes.

7.3. DATA

For carrying out the detailed logit analysis combined information of Tollygounge and Baranagar has been considered. As already mentioned, we will be carrying out the analysis of relation between energy consumption and disease burden first. For that, in all, one hundred and eighty observations were considered out of which ninety belong to Tollygounge and the rest ninety to Baranagar. Out of the diseases that the people suffer from, only Dyspnea or shortness of breath has been considered as it is the most common disease that results directly from exposure to pollution. The energy consumption in monetary terms has been considered, as the units are different for different energy used. The conversion into monetary unit has given us a uniform unit so that the consumption by the different income groups in all the areas can be compared.

The second analysis will show us how the change in exposure to pollution level will result in the occurrence of Dyspnea to the people of different income classes. In order to determine the exposure to the different polluted environment we have taken the number of hours of stay in office, in transport, and finally indoors. Air quality index for different environments have been calculated in the following way.

AQI = 2.82*SPM Exposure + 4.15*SO_2 Exposure + 0.02*CO Exposure + 1*NOx Exposure

Then the total exposure to the pollution is calculated by

TE = SUM ti*AQI where ti = hours of stay in different polluted environments.

The value of the direct total exposure has been calculated for all the surveyed household i.e., one hundred and eighty. Since smoking and sex are also considered to be significant variables in determining the pattern of diseases we have converted this information into binary choice where male and smokers are numbered one and female and non smokers are numbered zero. This is how the data have been developed and used for the major conclusion of the study.

7.4. Relation between Energy Consumption and Disease Burden

From the survey it has been observed that the proportion of people suffering from shortness of breath is highest for the lower income group and lowest for the high income group. The values for the HIG, MIG, and LIG are respectively as 0.22, 0.307 and 0.366. The average expenditures on energy consumption for all the three income classes are Rs. 1210.53, Rs.804.49 and Rs.558.79 respectively. These simple facts suggest that although the households belonging to the lower income class have the lowest expenditure on energy but the disease burden is highest for this group. On the other hand, the high income group has the highest expenditure on energy consumption but the disease burden is comparatively less. In order to see how a change in expenditure on energy consumption results in a change in probability of occurrence of Dyspnea or shortness of breath for different income class we have adopted a simple binary logistic regression.

The dependant variable is the disease of shortness of breath (SB) and the independent variable is the expenditure on energy (EXP). We have considered the left hand side variable as 1 for the disease occurrence and for the non occurrence it takes the value 0.

Higher Income Group

Logit (SB) = -2.48 + 0.0556EXP

The simplest way to get the value of the change in probability is in the following fashion. As an approximation, we can assume that the initial probability for explanatory variables at their mean values of sample observations to be equal to the sample probability (0.22). Then,

Change in Probability (SB) per Unit Change in EXP
= 0.0556*0.22* (1-0.22)
= 0.0095

It is a mean change in probability of having respiratory disease per unit. It tells us that if the expenditure on energy increases by one unit, then the chance that the individual will suffer from respiratory disease will increase by 0.0095.

Middle Income Group

Logit (SB) = -1.722 + 0.098EXP

As an approximation, we can assume that the initial probability for explanatory variables at their mean values of sample observations to be equal to the sample probability (0.307). Then,

Change in Probability (SB) per Unit Change in EXP

= 0.099*0.307*(1-0.307)

= 0.021

It tells us that if the expenditure on energy increases by one unit, then the chance that the individual will suffer from respiratory disease will increase by 0.021.

Lower Income Group

Logit (SB) = -0.60712 + 0.1417 EXP

As an approximation, we can assume that the initial probability for explanatory variables at their mean values of sample observations to be equal to the sample probability (0.366). Then,

Change in Probability (SB) per Unit Change in EXP

= 0.1417*0.366*(1-0.366)

= 0.032

Again if the expenditure on energy increases by one unit, then the chance that the individual will suffer from respiratory disease will increase by 0.032.

Interestingly we can conclude that an increase in the expenditure on energy has the greatest effect of disease burden on the lower income group and lowest for the higher income group.

Next we move on to see how a change in exposure to pollution in a change in probability of occurrence of Dyspnea across income classes.

7.5. Relationship between Disease Burden and Exposure to Pollution

Higher Income Group

The estimated dose response equation is as follows.

LOGIT (SB) = -0.272 + 0.25DE + 0.027SEX + 0.182SM

where DE = Daily Exposure to environmental pollution (incorporating SO_2, NO_x, SPM and CO);

SEX = 1 for Male, and 0 for Female,
Smoking = 1 for Yes and 0 for No;
SB =1 for Yes, and = 0 for No.

The estimated logit function is then interpreted as the probability function for suffering from SB. Literature suggests that daily exposure (DE), dummy for smoking (SM) (1 is yes, 0 if no) and sex are statistically- significant variable in deciding upon the probability of having Shortness of Breath.

We are interested in calculating the change in the value of the probability for a unit increase in direct exposure to pollution. The simplest way to get the value of the change in probability is in the following fashion. As an approximation, we can assume that the initial probability for explanatory variables at their mean values of sample observations to be equal to the sample probability (0.22).Then,

Change in Probability (SB) per Unit Change in DE to pollution = 0.25*0.22*(1-0.22)
= 0.042

This, however, is the mean change in probability of having respiratory disease per unit. Now what does the value of the probability signify. It tells us that if the Direct Exposure to pollution increases by one unit, then the chance that the individual will suffer from respiratory disease will increase by 0.042.

Middle Income Group

Now we consider the case of the MIG. The equation is given below:

LOGIT (SB) = -2.307 + 0.211DE + 0.698SEX + 2.09SM

We again calculate the change in probability for the middle income class. The sample probability for the occurrence of the respiratory disease is (0.307).

Change in Probability (SB) per Unit Change in DE to pollution = 0.211*0.307*(1-0.307)
= 0.0446

It is again a mean change in probability of having respiratory disease per unit. It tells us that if the Direct Exposure to pollution increases by one unit, then the chance that the individual will suffer from respiratory disease will increase by 0.0446.

Lower Income Group

Now we consider the case of the LIG. The equation is given below:

LOGIT (SB) = -.268 + 0.1128DE + 0.255SEX + 0.160SM

We again calculate the change in probability for the middle income class. The sample probability for the occurrence of the respiratory disease is (0.366).

Change in Probability (SB) per Unit Change in DE to pollution = 0.2128*0.366* (1-0.366)
= 0.049

It is again a mean change in probability of having respiratory disease per unit. It tells us that if the Direct Exposure to pollution increases by one unit, then the chance that the individual will suffer from respiratory disease will increase by 0.049.

From the above analysis we see that the change in the probability that an individual will suffer from the disease for a unit increase in exposure to pollution is highest for the lower income groups and lowest for the higher income group.

On the whole, we observed that increase in the expenditure on energy has a direct relation with the burden of diseases and that will affect mostly lower income group compared to other two. On the other hand, unit increase in exposure to pollution is also directly correlated with the individual suffering from shortness to breath. In this case, also lower income groups are much affected in exposure to pollution than higher and middle income groups.

It is alarming to note from the above economic analysis that in both the cases the relatively more sufferers are the people from the lower income group compared to that of the higher and middle income.

Chapter 8

CONCLUSION AND POLICY IMPLICATIONS

Air pollution kills millions of people annually and many more have to live with respiratory diseases that reduce their quality of life. The poor are more vulnerable, since they have to put up with poor housing and working conditions, and live closer to pollution sources (SIDA, 2006).

Air pollution is sourced from combustion of fossil fuel. Apart from the industry, households are also major consumer of commercial energy. The impact of air pollution on health, especially of the poorer section is a matter of serious concern to the academic communities, policy makers and practitioners in developing countries like India. There are many studies on air pollution and health, but no systematic and integrated assessment has been carried out to investigate the impacts of air pollution on health in general, and the poor in particular in India. The current book makes a modest attempt to this end.

The book estimates the industrial emissions of CO_2, SO_2 and NO_x in India during 1983-84 to 2003-4. The whole period has been divided for analysis into four sub periods (1983-84 to 1989-90, 1989-90 to 1993-94, 1993-94 to 1998-99, and 1998-99 to 2003-4) to investigate the changes in emissions and various sources of change in total CO_2, SO_2 and NO_x emissions in the Indian economy using input-output structural decomposition analysis (SDA). Further, it examines the contribution made by different income groups on the emissions of CO_2, SO_2 and NO_x in India. To assess the health impacts of the pollution on different income groups the study has done a micro survey for a metro city, Calcutta. The book evaluates the extent of deterioration in air quality, identifies the causal factors responsible for deterioration and finally assesses the impacts of deteriorated air quality on human health.

The study spreads over eight chapters. The first chapter compiles the basics of the air pollutants- features, sources, and impacts on health. It also shows the status of air pollution in megacities of the world. The interfaces of pollution, health impact, and poverty are also discussed in this chapter. Above all, an extensive Indian scenario has also been presented at the end of this chapter along with recently announced WHO guidelines.

Chapter 2 dealt with the literature survey based on the impact of air pollution on health. The survey is mainly focusing on the experience of the developing countries. The literature on India is also presented.

Chapter 3 calibrates the models to estimate the CO_2, SO_2 and NO_x emissions in India and the contributions made by the different income groups. Two models are developed. Model 1 estimates the generation of CO_2, SO_2 and NO_x emissions in the Indian economy for the period 1983-84 to 2003-4 by using Input-Output SDA approach. Model 2 which is an extension of model 1 derives the contribution of the total emissions by different income groups (higher, middle and lower categorized according to monthly consumption expenditure) adopting the same methodology.

Chapter 4 explains all the data information, sources and how these are processed. Two important informations are used --- input-output table from CSO and household consumption expenditure from NSSO.

Chapter 5 presents the analysis of the results. The final out come of the results shows that the industrial emissions of air pollutants have increased considerably in India during 1983–1984 to 2003-4 (11% per annum for CO_2, 12.67% per annum for SO_2, and 9.19% per annum for NO_x). The main factors for these increases are *changes in the final demand* throughout the period. *Changes in final demand* were increasing gradually in each phases of the study (100.82 mt of CO_2 in 1983-84 to 1989-90, 175.07 mt of CO_2 in 1989-90 to 1993-94, 213.60 mt of CO_2 in 1993-94 to 1998-99, and finally 304.56 mt of CO_2 in 1998-99 to 2003-4). The *change in intensity* has made a positive contribution in the first two periods i.e. it helped to increase emissions, but in the later half of the study period it showed a reduction. The *change in technology* always recorded a negative contribution except the first period. These effects helped to reduce the total changes in emission. The results of decomposition show that the economy is trying to adopt more efficient and less energy intensive technology, but the growth of final demand has rather worked in different direction, dwarfed the contribution of other factors, and aggravated the emission. Results of the contribution made by different income groups reveal that higher income groups are mostly responsible for such changes in emission. This has been due to their high level of consumption of energy. The middle income groups also contributed. However, lower group is a minor player. Considering factor wise contribution by different income groups, the higher income groups mostly influenced the intensity and final demand effect. The middle-income group's contribution is also considerable like the higher. The technology effect shows negative sign which helped to reduce the emission changes throughout, though with fluctuations. The overall assessment from the analysis of the results reveals that the contribution to the air pollution made by the lower income group is very insignificant while the higher income groups are major players for all emissions and almost for all factors.

Chapter 6 measures the air pollution effects on human health based on a micro survey. The micro study considers a metro city, Calcutta. The chapter outlines the features of the study site, overall environmental and health related problems in Calcutta along with a detailed survey methodology.

Chapter 7 developed the dose response function. It analyses the logit model for deriving the simple relationship between pollution, energy expenditure, and general health conditions. The micro study on Calcutta reveals that the people from lower income groups suffer more from the pollution led disease burden compared to people from higher income groups who are observed to spend more on energy. Further, people who live in poverty are those who are exposed to the worst environmental and health risks. Survey analysis supports that it is the poor who have the highest pollution led disease burden along with the greatest chances of suffering from the disease if they are exposed more to pollution. Our assessment reveals that higher income groups are almost responsible for generating emissions more than 75% and 20-22% is shared by middle income groups. But a very negligible amount has been contributed by lower income groups. Unfortunately, the lower income groups are suffering seriously due to pollution as evident from the survey analysis.

Estimates suggest that 25% and 33% of the global burden of diseases can be attributed to environmental factors (Kjellén, 2001). This proportion is larger in conditions of poverty, where more environmental hazards are present in the nearby living and working environment, and people have less capacity to protect themselves against exposure and effects of harmful and unpleasant pollutants. Pollution related health hazards are not uniform in all income groups. It affects the lower income groups more than the upper. Incidence of poverty is high in India and about one third of the population is below the poverty line. This section is mostly affected by environmental hazards.

In sum, we find that air pollution is growing in rapid pace and it has severe effects on human health leading to increased chronic bronchitis, respiratory problem, asthma, and cardiovascular problem in India. The study also suggests that health hazards are seriously caused by emissions of air pollution, primarily generated by the higher income groups in India. One of the biggest challenges at present is to tackle the generation of emission by the higher income groups along with the improvements of the health condition of the economy. It calls for proper integrated policy for the mitigation of the air pollution generation from the experiences of other countries.

On the basis of the results of our macro study the pollution measures can be targeted on most dominating factor responsible for generation of air pollution from fossil fuel combustion. We have seen that the final demand factor is one of the major components responsible for increasing emissions. It also shows that the higher income groups consume more energy and so also the major contributor of emissions. The relation between manmade emissions and climate change is claimed by the IPCC assessment report, 2007. The current study also supports that view. On the other hand, the micro part of our study shows that lower income groups are suffering relatively more from pollution led diseases than other two income groups. So a combined policy is needed at this stage to control the pollution as well as health problem. Our policy suggestion is targeted in that direction.

Consumption of energy by the final demand sector is enormous. Private consumption expenditure on energy and transport sector by the higher and middle income groups is relatively more than lower income groups so also emission generation.

So the emphasis on few basic broad policies like clean technology, increasing energy efficiency (conservation of energy) and fuel switching[19] in different sectors of the economy is required especially for transport[20].

---

19 One of the studies by GEO (2000) suggests that fuel switching policy in Asia Pacific countries could reduce the 2030 emission of Sulfur oxides to below the 1990 level, and limit the increase of nitrogen oxides to 40%. The study also shows that the technology to reduce environmental pressures in the region to sustainable levels is actually available. However, to make the necessary changes capital will be required.

20 The World Health Organization estimates that urban air pollution causes 800,000 premature deaths each year. Fossil fuels burnt by motor vehicles contribute 90 percent of urban air pollution, including lead, carbon monoxide, ozone and suspended particulate matter. We know that pollution generating from the transport sector is increasing in India. Although urban air pollution is a problem worldwide, vehicle emissions are particularly high in developing countries due to the prevalence of fuel-inefficient technologies and practices such as two stroke engines, high average vehicle age, poor vehicle maintenance, low-quality fuels, and severe traffic congestion. One of the major contributors to acid precipitation is the transport vehicle. The transportation sector also accounts for over 18 percent of global carbon dioxide emissions and is a significant contributor to global warming (Earth Trends, 2006). Currently, industrialized countries contribute roughly three times more transport-related greenhouse gas emissions than do developing countries. However, rapid motorization and population growth in the developing world are closing this gap. If China and India were to achieve per capita vehicle ownership levels comparable to those in the United States - roughly three cars per four people - the number of vehicles in the world would nearly triple. (Earth Trends, 2006).

Cleaner technologies should be made mandatory in industries in general, and new industries in particular so as to reduce pollution in the coming years. Industries are less willing to change their techniques of production once established. Hence, these industries should adopt cleaner technologies right from the begining. The technology for small-scale industries can be improved provided the large-scale manufacturers produce less polluting equipment.

Environmental audits for industry are necessary. Environmental impact assessments are often made at the start of the project but prescribed environmental management practices are often neglected once the project is underway. Regular audits are required to act as self-monitoring and enhance compliance to standards.

Priority is needed in case of conservation of energy that will play a significant role in alleviating the shortage of energy and in reducing environment pollution. The government has to ensure strict implementation of the energy conservation act which would promote rationalisation of energy consumption by especially targeting the transport sector.

An imposition of energy consumption tax on the higher income groups can be thought of by the policy makers. The tax rate can be decided on the basis of the volume of energy consumption by this group. The principle behind this is to put a tax on pollution. The more carbon dioxide one emits the more he pays in taxes.

There are various options that can reduce vehicular emissions. We are elaborating some of them.

Inter-fuel substitution can help to minimize the carbon emission problem to some extent. Fuel substitution in the household sector is highly required. Overall gaseous fuels such as CNG and LPG emit considerably less particulate matter than diesel vehicles not equipped with particulate traps. The other most popular method is *traffic demand management.* Strategies and measures can be introduced to regulate land use and traffic planning. The economic benefits of traffic management are significant. It reduces congestion and facilitates improved mobility. Improved traffic management confers environmental benefits because of the lower emission intensity of traffic by a)*lowering vehicle miles traveled (VMT),* through restraining transport demand, through increasing the role of urban public transit and increasing the occupancy of private vehicles; b)*increasing fuel efficiency*, so that less fuel is used (with lower pollution) for a given level of VMT; c)*reducing the rate of emissions per unit of fuel use ;d)* the *Vehicle Retirement and Scrappage programs*. The government should consider the introduction of clean fuel vehicle technology and especially fiscal incentives in this regard.

In this connection, we can mention the recently planned adaptation policies and mitigation technologies/policies announced by IPCC, 2007[21] for energy, human health, transport and industry.

---

[21] Planned adaptation policies and mitigation technologies, policies announced by IPCC-AR (2007) can be summarized as follows:

Planned adaptation by sector (IPCC-AR (2007))

Sector	Adaptation option/ strategy	Underlying policy framework	Key constraints and opportunities to implementation normal font =*constraint* italics = *opportunities*
Human health	Heat-health action plans; emergency medical services; improved climate-sensitive disease surveillance and control; safe water and improved sanitation	Public health policies that recognise climate risk; strengthened health services; regional and international cooperation	Limits to human tolerance (vulnerable groups); knowledge limitations; financial capacity; *upgraded health services; improved quality of life*
Transport	Realignment/relocation; design standards and planning for roads, rail, and other infrastructure to cope with warming and drainage	Integrating climate change considerations into national transport policy; investment in RandD for special situations, e.g. permafrost areas	Financial and technological barriers; availability of less vulnerable routes; *improved technologies and integration with key sectors (e.g. energy)*
Energy	Strengthening of overhead transmission and distribution infrastructure; underground cabling for utilities; energy efficiency; use of renewable sources; reduced dependence on single sources of energy	National energy policies, regulations, and fiscal and financial incentives to encourage use of alternative sources; incorporating climate change in design standards	Access to viable alternatives; financial and technological barriers; acceptance of new technologies; *stimulation of new technologies; use of local resources*
Sector	Key mitigation technologies and practices currently commercially available. *Key mitigation technologies and practices projected to be commercialised before 2030 (in italics)*	Policies, measures and instruments shown to be environmentally effective	Key constraints and opportunities normal font =*constraint* italics = *opportunities*
Transport	More fuel efficient vehicles; hybrid vehicles; cleaner diesel vehicles; biofuels; modal shifts from road transport to rail and public transport systems; non-motorised transport (cycling, walking); land use and transport planning*; *Second generation biofuels; higher efficiency aircraft; advanced electric and hybrid vehicles with more powerful and reliable batteries*	Mandatory fueleconomy, biofuel blending and CO_2 standards for road transport, Taxes on vehicle purchase, registration, use and motor fuels, road and parking pricing; Influence mobility needs through land use regulations, and infrastructure planning; Investment in attractive public transport facilities and non-motorised forms of transport	Partial coverage of vehicle fleet may limit effectiveness; Effectiveness may drop with higher incomes;

These options are already implemented by European countries, USA, and a number of Asian countries. Several regions of the world have been significantly tightening their motor vehicle regulations. Major developments during the past few years can be summarized.

Hungary introduced a retirement scheme for heavy-duty vehicles by offering US$3600 (in 1997 dollars) for replacement of an old bus with a new one that complied with the most recent emissions standards or for changing its engine (World Bank, 2000).

Another alternative is ULSD that contains 0.005% Sulfur content. ULSD has been tried in many European countries. It was introduced in Finland and Sweden in 1993 and has now captured 100% of the Scandinavian market. A study in New York city of the performance of ULSD buses, for instance, has shown that more than 90% reductions improvements in diesel technology, in particular the combination of ULSD and particulate matter traps, can indeed match the emission standards of CNG engines.

The EU has adopted Directives regarding light duty vehicle emissions and fuel quality that tighten standards significantly (2000 and 2005), broaden the scope of coverage, impose low sulfur requirements for diesel fuel and gasoline (Walsh,2003).

Energy Supply	Improved supply and distribution efficiency; fuel switching from coal to gas; nuclear power; renewable heat and power (hydropower, solar, wind, geothermal and bio-energy); combined heat and power; early applications of Carbon Dioxide Capture and Storage (CCS) (e.g. storage of removed CO_2 from natural gas); *CCS for gas, biomass and coal-fired electricity generating facilities; advanced nuclear power; advanced renewable energy, including tidal and wave energy, concentrating solar, and solar photovoltaics*	Reduction of fossil fuel subsidies; Taxes or carbon charges on fossil Fuels; Feed-in tariffs for renewable energy technologies; Renewable energy obligations; Producer subsidies	*May be appropriate to create markets for low emissions technologies*
Industry	More efficient end-use electrical equipment; heat and power recovery; material recycling and substitution; control of non-CO_2 gas emissions; and a wide array of process-specific technologies; *Advanced energy efficiency; CCS for cement, ammonia, and iron manufacture; inert electrodes for aluminium manufacture*	Provision of benchmark information;Performance standards; Subsidies, tax credits; tradable permit and voluntary agreement	*It is appropriate to stimulate technology uptake.* Stability of national policy important in view of international competitiveness; Predictable allocation mechanisms and stable price signals important for investments; Success factors include: clear targets, a baseline scenario, third party involvement in design and review and formal provisions of monitoring, close cooperation between government and industry

Source: IPCC AR4-2007.

The EU and the auto industry reached agreement on a voluntary commitment to reduce CO_2 emissions per kilometer driven by 25 per cent by about 2008 (Walsh, 2003).

The EU adopted the next phases of heavy duty standards – Euro 3, 4 and 5 - which will likely result in particulate and NO_x after treatment systems. The USEPA then tightened its heavy-duty engine emissions requirements with special focus on tighter PM and NO_x standards as well as low sulfur diesel fuel. It adopted a tightening of light duty vehicle standards closely modeled after the California LEV 2 standards (so called Tier 2) as well as tighter sulfur requirements in gasoline(Walsh, 2003). EPA has also adopted similar requirements for off road diesels and fuels (Walsh, 2003).

The California Air Resources Board (CARB) tightened the CO, HC, NO_x and PM requirements and also established the principles of fuel neutrality (diesel vehicles meet the same standards as gasoline fueled vehicles) and usage neutrality (light trucks and sports utility vehicles, used primarily as passenger cars, must meet the same standards as cars) (Walsh, 2003).

Japan tightened the gasoline fueled automobile standards and introduced very stringent diesel fueled vehicle requirements. The Ministry of International Trade and Industry (MITI) in Japan and Japanese industry reached agreement regarding lower CO_2 emissions from vehicles (Walsh, 2003).

The above stringent strategies are taken by the developed world. Developing countries also adopted some initiatives in this respect. We can mention some of them.

Alcohol biomass fuels for transport have played a major role in Brazil (Goldemberg et al., 1993) and plantations could provide significant bio fuels in many countries (Hall et al., 1993). Three-way catalytic converters can reduce the emissions of some pollutants. For gasoline vehicles, efficiently operated three way catalytic converters can reduce exhaust CO and hydrocarbon emissions by as much as 95% and NOx by over 75%.

In early 1990s, Chile used an effective Scrappage policy combined with tax incentives to remove the most polluting diesel buses from its urban transport fleet (World Bank,2000) .

A World Bank study concluded that a judicious use of gasoline tax could save the citizens of Mexico City US$110 million a year more than would an otherwise well-designed control program with no gasoline tax (Eskeland and Devarajan 1996).

China formally adopted the Euro 1 auto emissions standards and phased out the use of leaded gasoline across the entire country (Walsh, 2003).

New development plans for Beijing will change the transportation structure by encouraging public transportation. For in-use vehicles, the I/M (inspection and maintenance) program has employed ASM tests since early 2003 and the government has encouraged the retirement of high-emission vehicles. For new vehicles, Beijing introduced Euro 1, Euro 2 and Euro 3 emissions standards in early 1999, 2003, and 2005 respectively. At the same time, the fuel quality in Beijing was improved significantly, by banning lead and reducing sulfur among other changes. CNG and LPG were introduced in 1999 and are used in buses and taxis. Presently Beijing has the largest CNG bus fleet in the world with more than 2000 CNG buses. Beijing has also focused on fiscal incentives such as tax deductions for new vehicles meeting enhanced emission standards to encourage their sales. These strategies and measures have had an impact on the control of vehicular emissions. Despite the rapid increase of the vehicle population by 60% between 1998 and 2003, total vehicular emissions have not increased. With the enhancement of vehicular emission control, the air quality in Beijing is improving (Hao, Hu and Fu, 2006).

Taiwan adopted step 4 of its motorcycle control program, effectively banning two stroke motorcycles by 2003 (Walsh, 2003).

The experiences from the other countries of the world help us to derive the proper and suitable as well as feasible policies to reduce the emission from the transportation sector in India[22]. As we know that within the final demand factor, the private consumption expenditure is a major component, expenditure on transport being an important part. The number of vehicles is multiplying and so also the pollution load (the total number of registered motor vehicles increased from 1.86 million in 1971 to 67 million in 2003;ADB,2006). It indicates that almost all Indian cities, particularly metropolitan cities such as Delhi[23], Mumbai and Calcutta, are reeling under severe air pollution. In this context we briefly review the measures adopted in India for metropolitan cities.

In Delhi ----an integrated policy in the transport sector through vehicle scrappage, fuel substitution and financial incentive has already been implemented. The Supreme Court of India banned the sale of leaded gasoline in Delhi as of September 1999 as well as mandating that all new cars meet Euro 1 auto standards. Similar requirements were then phased in across the entire country in 2000. Delhi then adopted Euro 2 standards as of April 2000. Public transport in Delhi was amended by the Supreme Court of India to use Compressed Natural Gas (CNG) instead of diesel or petrol. Several options have been placed to implement that --- a) replacement of all pre-1990 autos and taxies are with new vehicles on clean fuels till March 1, 2000; b)financial incentives for replacement of all post 1990 autos and taxies with new vehicles on clean fuel by March 31, 2000; c)buses older than 8 years are not allowed to ply without Compressed Natural Gas (CNG) or other clean fuels by April 1, 2000. After that Delhi has the highest fraction of CNG-run public vehicles in the world and most of them were introduced within 20 months to reduce vehicular pollution. CNG powered vehicles emit 85% less NOx, 70% less reactive HCs and 74% less CO than similar gasoline powered vehicles[24]. A decreasing trend was found for SO_2 and CO concentrations, in comparison to those before the implementation of CNG (Ravindra et al., 2006).

The other cities in India are not as advanced as Delhi. From the study site description we have seen that auto emission is abnormally high in Calcutta which is dangerous to human health (WBPCB, 2001). In this tune strict measures like Delhi are needed for Calcutta to control the vehicular emissions.

In 1999 the Calcutta High Court passed an order which required all private or commercial vehicles in CMA to comply with Bharat Stage II (Euro II equivalent) standard and all three- and two-wheelers to comply with India 2000 standard (Euro I equivalent) (WBPCB, 2003; CSE, 2006). In 2001 stricter emissions standards were introduced for new four-wheeled vehicles and all other vehicles in CMA. Since 2001 there has been a ban on the distribution and sale of loose 2T oil (inert mixture of mobil oil and petrol) in CMA (Calcutta Metropolitan area) to address the problem with two-stroke two and three-wheelers.

---

[22] While the population in major cities of India has doubled during 1981–2001, the number of vehicle has gone up by six-and-a-half fold. (Reddy, 2007)

[23] There were 2.5 million vehicles registered in Delhi during 1996, while this number has reached 4.17 million in 2004 (Transport Department, 2004). Vehicular pollution accounts significantly to the total pollution generated in Delhi (Economic Survey of Delhi, 2003–04; Gurjar et al., 2004). The most common method of public transport in Delhi comprises buses, minibuses, taxies and three wheelers. Their consumption of diesel is higher as compared to gasoline than possibly anywhere in the Western World. Further, diesel constitutes two-third of the total fuel consumption in the transport sector (Aggarwal, 1999).

[24] (http://daq.state.nc.us/motor/cng/).

Apart from this some measures have been adopted in respect of industries in Calcutta. Still Calcutta's air in many places is really intolerable (detail in chapter 6)[25].

Besides the above mentioned strategies and measures several other important policy options are also in place.

As we know that, energy prices always play a crucial role in developing countries particularly while deriving policies. Prices should truly reflect the cost of using fossil fuels including the health cost from pollution and cost of environmental degradation. So a suitable and judicious energy price policy should be framed especially to influence the energy consumption of the higher income groups. The execution and target oriented price policy needs time. India has made some steps in this respect. It continuously increased prices of diesel, starting at the year 1995 at the lowest level (19 cents per litre) in the whole region of Asia and reaching 9 years later the highest diesel prices (62 cents per litre) in the developing part of Asia. The super gasoline price has also increased from 48 cents per litre in 1995 to 87 cents in 2004 (GTZ energy information, 2005). Inspite of the tough strategy taken by the government, the demand for fuel is still rising and so also emission, as evident from the current study. So the price policy is not the only solution in a country like India.

From the micro analysis we can suggest that more exhaustive local level epidemiological studies should be designed using monitors that measure CO, NO_x, SO_2, PM_{10} and the impacts of long-term exposure (especially for the poor people). It calls for better air quality policy and management.

Air quality policy should be viewed as an integral part of a set of much broader social policies, rather than as a stand-alone program. Already a revised air quality guideline (detail in Chapter 1) has been announced by the WHO (WHO, 2006) for worldwide implementation. Air quality is one of many basic needs. It increases productivity as well as directly enhancing human welfare through better health and an improved environment. Policy makers are expected to be more aware of the link between air quality and economic growth.

Policies to increase economic growth and development must include provisions to improve air quality. Already we discussed about the policies that will reduce emission generating from industrial and household activity. Those policies will be implied as air quality policies indirectly. As for example, investment in energy efficiency, energy conserving technologies, and reducing or removing energy subsidies can reduce pollution while conserving financial resources. These approaches will produce cleaner air at little or no marginal cost for the air quality improvement.

To strengthen a preventive and integrated approach to AQM[26] (air quality management), the use of economic instruments will have to be widely applied in managing stationary and mobile air pollution. Singapore led the way by doing one of the first countries to introduce congestion charges, which has been followed by Trondheim, Norway and London, U.K.

---

[25] In 2001, under a notification by the West Bengal Pollution Control Board (WBPCB), the standards for PM emissions were set at 150 mg/nm^3 for all boilers irrespective of their steam generation capacity, for all ceramic kilns irrespective of the nature of kiln, and for all cast iron foundries (cupola furnaces), irrespective of their metal melting capacity, for all rolling mills (WBDoE, 2003).

[26] Air pollution management requires the establishment of national or local laws, and institutions to assess pollution levels and enforce the laws. Reliable information on a city's air pollution sources and its actual air quality are prerequisite to air quality management. Institutions capable of collecting and analyzing such data and information must be developed. Since there are competing options for air pollution control, many of them requiring substantial capital investments, it is crucial for management to be cost effective.

Congestion charging has proved to be an effective measure to reduce private vehicle use, improve public transport and reduce emissions and congestions(BBC news,2004; BBC 2006).

India while experiencing rapidly rising pollution levels, are in the early stages of development of pollution abatement policies. Already in India several acts have been passed and implemented along with air quality improvement strategy[27]. However, our study suggests that the status of air pollution in Indian urban areas is not in a good shape.

The key requirement at this stage for India is building capacity formulation, assessment, selection, and implementation of air pollution control policies as quickly as possible. To achieve the target, policy makers of our country should address the following questions.

What type and how much of different pollutants are emitted? Are they increasing or decreasing and to what degree? What factors are responsible for these changes? What health and other costs do they impose on residents? What is the level of emission and ambient air quality standards that need to be established to avoid such damages? What policy instruments can be identified that can meet the air quality standards? What analytical and empirical models can be developed and used for evaluating and choosing from among these policy instruments? To deal with these questions there is a need to improve the data collection, assessment and analysis so that decisions can be based on sound scientific information. It is to be mentioned that the essence of policy formulation should be treated as a dynamic and on-going continuous process of monitoring, analyzing and reviewing policies (Lakhsmanan, 2004).

We have seen from the study that populations living in cities with high levels of air pollution in developing countries experience greater adverse effects on health. On the other hand, epidemiological studies in developed countries (detail in chapter 2) suggest that air pollution affects both mortality and morbidity rates significantly, and generates high social costs associated with premature death and a decrease in the quality of life. In these developed countries, considerable amount of resources are used for clean-ups and remedial actions, safety standards, and regulations along with enforcement are becoming stricter. In contrast, developing countries generally do not have a luxury of diverting substantial resources to pollution abatement programs, and thus find themselves in a dilemma of trade-offs of rapid economic growth and environmental protection. This inhibits the adequate allocation of resources for addressing the environmental problems.

Protecting health by reducing high levels of pollution is one of many challenges. Many developing countries are grappling with the challenge of addressing hazardous levels of air pollution especially in urban areas. The valuation of illness and premature death due to air pollution in the context of India is an issue that needs more attention from the national and international community of environmental economists and institutions. Successful programs to protect health require a broader base of policies than do traditional regulations or complex

---

[27] The 1981 Air (Prevention and Control of Pollution) Act amended in 1987 and the 1982 Air (Prevention and Control of Pollution) Rules provide the legal basis for air pollution abatement measures(MOEF, 2002). The Air Act provides for the control and abatement of air pollution while the Air Rules define the procedures of the Central Pollution Control Board (CPCB) and State Pollution Control Boards (SPCBs). Additional key legislation and programmes relating to air pollution include: (i) The Environment (Protection) Act, 1986; (ii) The National Policy Statement on Abatement of Pollution (1992); and (iii) The Environment Action Programme(1993)(Male Declaration,2000). National ambient air quality standards for SO_2, NO_2, CO, TSP, PM_{10} and lead with short-term (24-hours) and long-term (annual) limits have been set for industrial, residential (including rural and other areas) and sensitive areas in order of decreasing values for set limits(Schwela et al., 2006).

economic incentives. In a country like India, where resources are very scarce relative to a variety of development objectives, we should give special importance to carefully set air pollution control policies and priorities considering the health impacts and anticipated benefits. To this end more information is needed to assess the health effects of air pollution in India and other developing countries and efforts should be targeted to increase the number of epidemiological studies.

ACKNOWLEDGEMENTS

This book is partially an outcome of the research project funded by the Indian Council of Social Science Research (ICSSR), New Delhi during the period 2003-6 at the Centre for Development and Environment Policy, Indian Institute of Management Calcutta, India.

I gratefully acknowledge Indian Council of Social Science Research (ICSSR), New Delhi for the financial support to carry out this academic research. I wish to acknowledge Indian Institute of Management Calcutta for providing necessary facilities during the project. In this context, I express my sincere thanks to the Project Director, Prof. Jayanta Bandyopadhyay (Head, Centre for Development & Environmental Policy, Indian Institute of Management Calcutta, and President INSEE), for his continuous support and encouragement.

I also acknowledge Research scholars (Souvik Bhattacharjee, Sukanya Bose and Debabrata Mitra) who have worked in the project for a short period.

I express my heartfelt gratitude to Prof. Debesh Chakraborty (Department of Economics, Jadavpur University, Calcutta, India) for having ingrained me in this subject and encouraged me to give a concrete shape of this work. He has taken great interest and provided immense support and guidance while preparing the manuscript and refining the text. He has remained a source of inspiration throughout my current work.

This study would not have been completed without assistance from various government organisations and research institutes. In this respect thanks are due to National Sample Survey Organisation (NSSO) for providing necessary data. I also gratefully acknowledge CSO (Central Statistical Organisation, New Delhi) especially Dr. R.P.Kohli (Addl. Director General, CSO) for providing Input -Output tables and necessary inputs for the study.

I wish to acknowledge Mr. A. Rahman (Moonlit Computer, Santoshpur, Calcutta) and Shampa Kundu (PhD.Student, Economics Department, Jadavpur University, Calcutta) for processing NSSO data and other computer assistance.

I am deeply indebted to Prof. Paul Thomassin (Department of Agricultural Economics, McGill University, Montreal, Canada) for his constructive ideas and above all the financial support to complete the monograph.

This book would not have appeared without the support and facilities provided by McGill University in general, and Department of Agricultural Economics in particular.

Finally, my mother provided me with ample space and love, and her patience carried me through the work.

APPENDICES

APPENDIX A1

Estimated Total Deaths ('000), by Major Causes in Asian Developing Countries, 2002

		Afghanistan	Bangladesh	Cambodia	China	India	Indonesia	Iran	Iraq
Population ('000)		*22,930*	*143,809*	*13,810*	*1,302,307*	*1,049,550*	*217,131*	*68,070*	*24,510*
All Causes		**484.5**	**1,106.8**	**160.5**	**9,135.5**	**10,378.5**	**1,626.1**	**384.5**	**213.2**
Respiratory infections		**56.7**	**125.1**	**8.9**	**291.8**	**1,123.1**	**106.3**	**7.0**	**23.6**
1.	Lower respiratory infections	55.9	124.1	8.3	268.8	1,107.9	104.9	6.9	22.4
2.	Upper respiratory infections	0.8	1.0	0.5	23.0	14.7	1.3	0.1	1.2
3.	Otitis media	0.0	0.0	0.0	-	0.5	0.1	-	0.0
Perinatal conditions		**62.0**	**90.1**	**11.5**	**272.7**	**762.1**	**73.3**	**18.0**	**14.3**
1.	Low birth weight	34.2	61.2	4.3	71.0	542.5	45.3	8.2	7.3
2.	Birth asphyxia and birth trauma	17.2	16.0	4.0	96.9	129.3	17.7	4.0	3.6
Cardiovascular diseases		**67.6**	**253.4**	**21.8**	**3,001.3**	**2,810.0**	**468.7**	**157.6**	**45.7**
1.	Rheumatic heart disease	1.9	10.3	0.6	97.2	103.9	11.7	1.1	0.7
2.	Hypertensive heart disease	7.5	14.6	3.5	219.0	49.7	39.4	11.5	4.3
3.	Ischaemic heart disease	33.2	130.0	7.6	702.9	1,531.5	220.4	82.0	22.0
4.	Cerebrovascular disease	11.5	64.5	6.0	1,652.9	771.1	123.7	31.8	8.3
5.	Inflammatory heart diseases	2.3	5.0	0.3	65.8	57.8	8.2	7.4	2.1
Respiratory diseases		**8.6**	**57.1**	**5.3**	**1,432.5**	**609.5**	**109.7**	**15.4**	**6.1**
1.	Chronic obstructive pulmonary disease	3.3	39.5	1.9	1,283.7	485.8	73.1	9.0	2.6
2.	Asthma	2.3	8.5	1.5	16.1	57.1	15.0	0.0	0.4

Estimated Total Deaths ('000), by Major Causes in Asian Developing Countries, 2002 (Continued)

		Israel	Kazakhstan	Malaysia	Maldives	Nepal	Pakistan	Sri Lanka	Thailand	Viet Nam
Population ('000)		*6,304*	*15,469*	*23,965*	*309*	*24,609*	*149,911*	*18,910*	*62,193*	*80,278*
All Causes		**35.4**	**184.1**	**119.2**	**2.1**	**233.3**	**1,386.4**	**145.5**	**419.1**	**515.8**
Respiratory infections		**0.8**	**5.2**	**7.2**	**0.2**	**24.2**	**167.6**	**6.6**	**12.8**	**28.1**
1.	Lower respiratory infections	0.8	4.4	6.9	0.2	23.8	164.6	5.6	12.1	26.7
2.	Upper respiratory infections	0.0	0.8	0.2	0.0	0.3	2.8	1.0	0.5	1.3
3.	Otitis media	0.0	0.0	0.0	0.0	0.0	0.2	0.0	0.2	0.1
Perinatal conditions		**0.4**	**2.9**	**2.1**	**0.2**	**24.5**	**114.0**	**2.3**	**6.8**	**18.9**
1.	Low birth weight	0.2	0.9	0.8	0.1	15.0	81.2	0.9	3.2	11.1
2.	Birth asphyxia and birth trauma	0.1	0.7	0.8	0.0	5.0	19.0	0.7	3.1	4.6
Cardiovascular diseases		**11.1**	**96.4**	**35.7**	**0.6**	**49.9**	**286.8**	**47.7**	**83.6**	**160.0**
1.	Rheumatic heart disease	0.2	0.9	0.5	0.0	1.6	11.6	0.2	0.5	4.2
2.	Hypertensive heart disease	0.6	5.6	5.6	0.1	6.2	5.1	9.6	7.2	7.5
3.	Ischaemic heart disease	5.7	51.9	13.4	0.3	23.3	154.3	16.3	28.4	66.2
4.	Cerebrovascular disease	2.2	26.9	10.2	0.2	12.0	78.5	13.3	24.8	58.3
5.	Inflammatory heart diseases	0.2	2.6	0.5	0.0	0.8	6.2	0.0	1.4	3.2
Respiratory diseases		**1.8**	**7.9**	**9.7**	**0.1**	**11.1**	**62.5**	**15.6**	**27.5**	**51.3**
1.	Chronic obstructive pulmonary disease	1.0	5.2	6.0	0.1	6.6	48.6	9.5	17.6	41.9
2.	Asthma	0.2	1.0	1.4	0.0	2.2	5.7	2.9	2.5	2.5

Source: Compiled by the author from WHO,2002 http://www.who.int/healthinfo/bodestimates/en/index.html

Appendix A2

Estimated Total DALYs ('000), by Major Causes in Asian Developing Countries, 2002

		Afghanistan	Bangladesh	Cambodia	China	Indonesia	Iran	Iraq	Israel
Population ('000)		*22,930*	*143,809*	*13,810*	*1,302,307*	*217,131*	*68,070*	*24,510*	*6,304*
All Causes		**17,011**	**36,972**	**5,310**	**200,273**	**46,385**	**12,679**	**8,279**	**659**
Respiratory infections		**2,067**	**3,239**	**248**	**6,113**	**1,413**	**153**	**847**	**6**
1.	Lower respiratory infections	2,031	3,177	227	5,383	1,338	132	799	4
2.	Upper respiratory infections	28	19	16	490	22	4	41	0
3.	Otitis media	8	43	6	240	53	18	8	1
Cardiovascular diseases		**1,149**	**3,128**	**326**	**25,233**	**5,191**	**1,517**	**585**	**69**
1.	Rheumatic heart disease	59	234	18	1,401	229	24	19	1
2.	Hypertensive heart disease	98	180	43	1,660	389	87	41	3
3.	Ischaemic heart disease	469	1,442	105	5,130	2,313	726	230	30
4.	Cerebrovascular disease	208	684	87	14,555	1,311	309	110	17
5.	Inflammatory heart diseases	62	118	7	568	145	87	35	3
Respiratory diseases		**261**	**1,222**	**116**	**12,603**	**1,936**	**465**	**205**	**43**
1.	Chronic obstructive pulmonary disease	53	507	25	9,144	1,137	211	44	14
2.	Asthma	103	441	45	2,254	411	120	55	12

		Kazakhstan	Malaysia	Maldives	Nepal	Pakistan	Sri Lanka	Thailand	Viet Nam
Population ('000)		*15,469*	*23,965*	*309*	*24,609*	*149,911*	*18,910*	*62,193*	*80,278*
All Causes		**3,752**	**3,505**	**60**	**7,469**	**44,821**	**3,500**	**12,755**	**13,360**
Respiratory infections		**134**	**86**	**4**	**583**	**4,698**	**80**	**331**	**460**
1.	Lower respiratory infections	109	76	3	570	4,552	69	300	418
2.	Upper respiratory infections	23	3	0	5	94	8	15	19
3.	Otitis media	2	7	0	8	52	4	16	23
Cardiovascular diseases		**871**	**370**	**6**	**614**	**3,091**	**410**	**975**	**1,389**
1.	Rheumatic heart disease	18	10	0	39	238	3	9	61
2.	Hypertensive heart disease	56	51	1	73	63	74	56	58
3.	Ischaemic heart disease	409	133	3	259	1,493	144	347	564
4.	Cerebrovascular disease	249	105	1	134	730	113	258	455
5.	Inflammatory heart diseases	37	9	0	17	130	1	27	44
Respiratory diseases		**175**	**156**	**3**	**222**	**1,234**	**231**	**453**	**565**
1.	Chronic obstructive pulmonary disease	97	59	1	82	566	131	152	316
2.	Asthma	21	63	1	82	410	58	145	149

Source: Compiled by the author from WHO, 2002 http://www.who.int/healthinfo/bodestimates/en/index.html

APPENDIX B

B1. Aggregation Scheme of NSSO-IO Table

Sec. no.	Name of the aggregated sector	Sectors in original input-output table of 1983-84,1989-90, 1993-94, 1998-99	NSSO data (code no.)
1	Agricultural crops	Paddy (1),wheat(2), jowar (3), bajra (4),maize (5), gram (6),pulses (7), sugarcane (8), groundnut (9), jute (10),cotton (11), tea (12), coffee (13), rubber (14),coconut (15), tobacco (16), other crops (17)	(101-108)+(110-118)+(120-122)+(130-142)+(150-151) +(160-161)+(170-171)+(190-191)+(200-208)+(210-228)+(240-268)+(290-298)+(300-308)+(310-328)+(340-358)+(360-368)
2	Animal husbandry	Milk and milk products (18),Animal services(agricultural)(19), Other livestock products (20)	(270-288)+(370-378)+(390-398)
3	Forestry and logging	Forestry and logging(21)	
4	Fishing	Fishing (22)	
5	Coal and lignite	Coal and lignite (23)	(460-478)
6	Crude petroleum, natural gas	Crude petroleum, natural gas (24)	
7	Metallic and non-metallic minerals	Iron ore (25),Manganese ore (26),Bauxite (27),Copper ore (28),Other metallic minerals (29),Lime Stone (30),Mica (31),Other non metallic minerals (32)	
8	Food and Food products	Sugar (33), Khandsari, boora (34), Hydrogenated oil (vanaspati)(35),Edible oils other than vanaspati(36),Tea and coffee processing (37), Miscellaneous food products (38)	
9	Beverages	Beverages (39)	(409-419)
10	Tobacco products	Tobacco products(40)	(440-448)+(450-458)
11	Cotton textiles	Khadi, cotton textiles (handlooms) (41), Cotton textiles(42)	(480-491)+(493+501)
12	Wool, Silk and Synthetic fibre textiles	Woolen textiles (43),Silk textiles (44), Art silk, synthetic fiber textiles(45)	(502+492+494)
13	Jute, hemp, mesta textiles	Jute, hemp, mesta textiles(46)	
14	Other textile products	Carpet weaving(47),Readymade garments (48),Miscellaneous textile products(49)	(495+496+497+500)

B1. (Continued).

Sec. no.	Name of the aggregated sector	Sectors in original input-output table of 1983-84,1989-90, 1993-94, 1998-99	NSSO data (code no.)
15	Wood and wood products including furniture etc.	Furniture and fixtures-Wooden(50), Wood and wood products(51)	(690-708)
16	Paper, printing andpublishing	Paper, paper prods. and newsprint (52), Printing and publishing (53)	
17	Leather and leather products	Leather footwear(54),Leather and leather products(55)	(510-518)
18	Rubber products	Rubber products(56)	
19	Plastic products	Plastic products(57)	555
20	Petroleum products	Petroleum products(58)	
21	Coal tar products	Coal tar products(59)	
22	Inorganic heavy chemicals	Inorganic heavy chemicals(60)	
23	Organic heavy chemicals	Organic heavy chemicals(61)	
24	Fertilizers	Fertilizers(62)	
25	Paints, varnishes and lacquers	Paints, varnishes and lacquers(64)	
26	Other chemicals	Pesticides(63),Drugs and medicines (65) Soaps, cosmetics and glycerin (66), Synthetic fibers, resin(67),Other chemicals(68)	(540+557+568)
27	Cement	Cement(70)	
28	Clay and non-metallic mineral products	Structural clay products(69),Other non-metallic mineral products(71)	553
29	Iron and Steel	Iron, steel and ferro alloys(72),Iron and steel casting and forging(73),Iron and steel foundries(74)	
30	Non-ferrous basic metals	Non-ferrous basic metals(75)	
31	Misc. metal products	Hand tools, hardware(76),Miscellaneous metal products(77)	(730-739)+(740-748)
32	Tractors and agricultural mach.	Tractors and agricultural mach.(78)	
33	Industrial machinery(F and T)	Industrial machinery(F and T)(79)	
34	Other Machinery	Industrial machinery (others) (80), Machine tools(81),Office computing machines(82),Other non-electrical machinery(83)	

B1. (Continued).

Sec. no.	Name of the aggregated sector	Sectors in original input-output table of 1983-84,1989-90, 1993-94, 1998-99	NSSO data (code no.)
35	Electrical and Electronic appliance	Electrical industrial Machinery (84), Electrical wires and cables (85), Batteries (86),Electrical appliances (87), Communication equipments(88),Other electrical Machinery(89),Electronic equipments(incl.TV)(90)	769+729
36	Rail equipments	Rail equipments(92)	(529+539+549+569)
37	Other transport equipments	Ships and boats(91),Motor vehicles (93) ,Motor cycles and scooters(94),Bicycles, cycle-rickshaw(95),Other transport equipments(96)	(770-778)
38	Misc. Manufacturing Ind.	Watches and clocks(97),Miscellaneous manufacturing(98)	579+(790-798)+808
39	Construction	Construction(99)	
40	Electricity	Electricity(100)	
41	Gas and Water	Gas(101),Water supply(102)	
42	Railway transport services	Railway transport services(103)	
43	Other transport services	Other transport services(104)	
44	Storage and warehousing	Storage and warehousing(105)	
45	Communication	Communication(106)	
46	Medical and health	Medical and health(113),	479+678
47	Misc. Services	Trade(107),Hotels and restaurants (108), Banking(109),Insurance(110),Ownership of dwellings(111),Education and Research(112),Other services (114), Public Administration(115)	659+649+629

B2 Aggregation Scheme of NSSO-IO Table(2003-4)

Sec. no.	Name of the aggregated sector	Sectors in original input-output table of 2003-4	NSSO data (code no.)
1	Agricultural crops	Paddy (1),wheat(2), jowar (3), bajra (4),maize (5), gram (6),pulses (7), sugarcane (8), groundnut (9), coconut(10),other oil seeds(11),jute (12),cotton (13), tea (14), coffee (15), rubber (16) , tobacco (17), fruits(18),vegetables(19), other crops (20)	(101-108)+(110-118)+(120-122)+(130-142)+(150-151) +(160-161)+(170-171)+(190-191)+(200-208)+(210-228)+(240-268)+(290-298)+(300-308)+(310-328)+(340-358)+(360-368)
2	Animal husbandry	Milk and milk products (21),Animal services(agricultural) (22), poultry and eggs(23), Other livestock products and gobar gas(24)	(270-288)+(370-378)+(390-398)
3	Forestry and logging	Forestry and logging(25)	

B2. (Continued).

Sec. no.	Name of the aggregated sector	Sectors in original input-output table of 2003-4	NSSO data (code no.)
4	Fishing	Fishing (26)	
5	Coal and lignite	Coal and lignite (27)	(460-478)
6	Crude petroleum, natural gas	natural gas (28), Crude petroleum(29)	
7	Metallic and non-metallic minerals	Iron ore (30),Manganese ore (31),Bauxite (32),Copper ore (33),Other metallic minerals (34),Lime Stone (35),Mica (36),Other non metallic minerals (37)	
8	Food and Food products	Sugar (38), Khandsari, boora (39), Hydrogenated oil (vanaspati)(40),Edible oils other than vanaspati(41),Tea and coffee processing (42), Miscellaneous food products (43)	
9	Beverages	Beverages (44)	(409-419)
10	Tobacco products	Tobacco products(45)	(440-448)+(450-458)
11	Cotton textiles	Khadi, cotton textiles (handlooms) (46), Cotton textiles(47)	(480-491)+(493+501)
12	Wool, Silk and Synthetic fibre textiles	Woolen textiles (48),Silk textiles (49), Art silk, synthetic fiber textiles(50)	(502+492+494)
13	Jute, hemp, mesta textiles	Jute, hemp, mesta textiles(51)	
14	Other textile products	Carpet weaving(52),Readymade garments (53),Miscellaneous textile products(54)	(495+496+497+500)
15	Wood and wood products including furniture etc.	Furniture and fixtures-Wooden(55), Wood and wood products(56)	(690-708)
16	Paper, printing andpublishing	Paper, paper prods. and newsprint (57), Printing and publishing (58)	
17	Leather and leather products	Leather footwear(59),Leather and leather products(60)	(510-518)
18	Rubber products	Rubber products(61)	
19	Plastic products	Plastic products(62)	555
20	Petroleum products	Petroleum products(63)	

B2. (Continued).

Sec. no.	Name of the aggregated sector	Sectors in original input-output table of 2003-4	NSSO data (code no.)
21	Coal tar products	Coal tar products(64)	
22	Inorganic heavy chemicals	Inorganic heavy chemicals(65)	
23	Organic heavy chemicals	Organic heavy chemicals(66)	
24	Fertilizers	Fertilizers(67)	
25	Paints, varnishes and lacquers	Paints, varnishes and lacquers(69)	
26	Other chemicals	Pesticides(68),Drugs and medicines (70) Soaps, cosmetics and glycerin (71), Synthetic fibers, resin(72),Other chemicals(73)	(540+557+568)
27	Cement	Cement(75)	
28	Clay and non-metallic mineral products	Structural clay products(69),Other non-metallic mineral products(76)	553
29	Iron and Steel	Iron, steel and ferro alloys(77),Iron and steel casting and forging(78),Iron and steel foundries(79)	
30	Non-ferrous basic metals	Non-ferrous basic metals(80)	
31	Misc. metal products	Hand tools, hardware(81),Miscellaneous metal products(82)	(730-739)+(740-748)
32	Tractors and agricultural mach.	Tractors and agricultural mach.(83)	
33	Industrial machinery(F and T)	Industrial machinery(F and T)(84)	
34	Other Machinery	Industrial machinery (others) (85), Machine tools(86), Other non-electrical machinery(87)	
35	Electrical and Electronic appliance	Electrical industrial Machinery (88), Electrical wires and cables (89), Batteries (90),Electrical appliances (91), Communication equipments(92),Other electrical Machinery(93),Electronic equipments(incl.TV)(94)	769+729
36	Rail equipments	Rail equipments(96)	(529+539+549+569)
37	Other transport equipments	Ships and boats(95),Motor vehicles (97) ,Motor cycles and scooters(98),Bicycles, cycle-rickshaw(99),Other transport equipments(100)	(770-778)
38	Misc. Manufacturing Ind.	Watches and clocks(101),Medical precision and optical instrument (102), Jems and Jewelry(103), Space craft and air craft(104), Miscellaneous manufacturing(105)	579+(790-798)+808
39	Construction	Construction(106)	
40	Electricity	Electricity(107)	
41	Gas and Water	Water supply(108)	

B2. (Continued)

Sec. no.	Name of the aggregated sector	Sectors in original input-output table of 2003-4	NSSO data (code no.)
42	Railway transport services	Railway transport services(109)	
43	Other transport services	Land transport incl.via pipeline(110), water transport(111), air transport (112), supporting and auxiliary transport activities(113)	
44	Storage and warehousing	Storage and warehousing(114)	
45	Communication	Communication(115)	
46	Medical and health	Medical and health(122),	479+678
47	Misc. Services	Trade(116),Hotels and restaurants (117), Banking(118),Insurance(119),Ownership of dwellings(120),Education and Research(121),Business services (123), computer and related activities (124), legal service(125), real estate services(126), renting of machinery and equipment (127), O. comm. social and personal services(128), Other services (129), Public Administration(130)	659+649+629

APPENDIX B3

NSSO Code List

Sl. No.	Item	Code	Sl. No.	PULSES AND PULSE PRODUCT	
1	paddy	101	50	arhar (PDS)	210
2	rice (PDS)	102	51	arhar (others)	211
3	rice (others)	103	52	gram (PDS)	212
4	chira	104	53	gram (others)	213
5	khoi/lawa	105	54	moong (PDS)	214
6	muri	106	55	moong (others)	215
7	other rice products	108	56	masur (PDS)	216
8	rice (1-7)	109	57	masur (others)	217
9	wheat(PDS)	110	58	urd (PDS)	218
10	wheat(others)	111	59	urd (others)	219
11	atta (PDS)	112	60	khesari (PDS)	220
12	atta (others)	113	61	khesari (others)	221
13	maida	114	62	peas	222
14	suji	115	63	soyabin	223
15	sewai/noodles	116	64	other pulses	224
16	bread	117	65	other pulses products	225
17	other wheat products	118	66	pulses and pulses products (50-65)	229
18	wheat (9-17)	119	MILK AND MILK PRODUCT		
19	jowar (PDS)	120	67	milk liquid (litre)	230
20	jowar (others)	121	68	baby food	231
21	jowar products	122	69	milk powder	232
22	jowar (19-21)	129	70	curd	233
23	bajra (PDS)	130	71	ghee	234
24	bajra (others)	131	72	butter	235
25	bajra products	132	73	ice cream	236
26	bajra (23-25)	139	74	other milk products	237
27	maize (PDS)	140	75	milk and milk products (67-74)	239
28	maize (others)	141	EDIBLE OIL		
29	maize products	142	76	vanaspati (PDS)	240
30	maize (27-30)	149	77	vanaspati (others)	241
31	barley	150	78	mustard oil (PDS)	242
32	barley products	151	79	mustard oil (others)	243
33	barley (31-32)	159	80	ground nut oil (PDS)	244
34	small millets	160	81	ground nut oil (others)	245
35	small millets products	161	82	coconut oil (PDS)	246
36	small millets (34-35)	169	83	coconut oil (others)	247
37	ragi	170	84	gingelly oil (PDS)	248
38	ragi products	171	85	gingelly oil(others)	249
39	ragi (37-38)	179	86	linseed oil (PDS)	250
40	TOTAL CEREAL (8+18+22+26+30+33+36)	189	87	linseed oil (others)	251
GRAM			88	refined oil (PDS)	252
41	gram (whole grain)	190	89	refined oil (others)	253
42	gram products	191	90	palm oil (PDS)	254
43	TOTAL GRAM (42-43)	199	91	palm oil (others)	255
CEREALS SUBSTITUTES			92	repseed (PDS)	256
44	tapioca /sago	200	93	repseed(others)	257
45	tapioca green	201	94	oil seed	258

B3. (Continued).

46	jackfruit seed	**202**	95	edible oil	**259**
47	other cereal substitute	**203**	96	edible oil (76-97)	**269**
48	cereal substitute (44-47)	**209**			
MEAT, EGG, FISH			FRESH FRUITS		
97	goat meat	**270**	146	banana	**340**
98	mutton	**271**	147	jack fruit	**341**
99	beaf	**272**	148	water melon	**342**
100	pork	**273**	149	pine apple	**343**
101	buffalo meat	**274**	150	coconut	**344**
102	other meat	**275**	151	guava	**345**
103	chicken	**276**	152	singara	**346**
104	other birds	**277**	153	orange,mausami	**347**
105	egg	**278**	154	mango	**348**
106	fish (fresh)	**279**	155	kharbooza	**349**
107	fish (dry)	**280**	156	pears (naspati)	**350**
108	fish (canned)	**281**	157	berries	**351**
109	others	**282**	158	leechi	**352**
110	egg,meat,fish (97-109)	**289**	159	apple	**353**
VEGETABLES			160	grapes	**354**
111	potato	**290**	161	other fresh fruits	**355**
112	onion	**291**	162	fresh fruits (146-161)	**359**
113	radish	**292**	FRUITS AND NUTS		
114	carrot	**293**	163	coconut (copra)	**360**
115	turnip	**294**	164	ground nut	**361**
116	beet	**295**	165	cashew nut	**362**
117	sweet potato	**296**	166	walnut	**363**
118	arum	**297**	167	other nuts	**364**
119	other root veg	**298**	168	raisin (kismis)	**365**
120	root vegetable (111-119)	**299**	169	other dry fruits	**366**
121	pumkin	**300**	170	dry fruits (163-170)	**369**
122	gourd	**301**	SUGAR		
123	bitter gourd	**302**	171	sugar crystal (PDS)	**370**
124	cucumber	**303**	172	sugar crystal (others)	**371**
125	parwal/patal	**304**	173	khandsari	**372**
126	jhinga	**305**	174	gur(cane)	**373**
127	snake gourd	**306**	175	gur (others)	**374**
128	other gourd	**307**	176	sugar candy (misri)	**375**
129	gourd (121-128)	**309**	177	honey	**376**
130	cauliflower	**310**	178	sugar (others)	**377**
131	cabbage	**311**	179	sugar all (171-178)	**379**
132	lady's finger	**312**	SALT		
133	palak	**313**	180	sea salt	**380**
134	other leafy vegetables	**314**	181	other salt	**381**
135	fresh bins/barbati	**315**	182	salt all (180-181)	**389**
136	tomato	**316**	SPICES		
137	peas	**317**	183	turmeric	**390**
138	chilli	**318**	184	black papper	**391**
139	capsicum	**319**	185	dry chillies	**392**
140	plantatain(green)	**320**	186	garlic	**393**
141	jack fruit (green)	**321**	187	tamarind	**394**
142	lemon	**322**	188	ginger	**395**

B3. (Continued).

143	other vegetables	323	189	curry powder	396
144	other vegetables(130-143)	329	190	other spices	397
145	all vegetable (120+129+144)	339	191	spices all (183-190)	399
BEVERAGES, REFRESHMENTS AND PROCESSED FOOD			241	kerosene (others)	465
192	tea (liquid)	400	242	matches	466
193	tea (leaf)	401	243	coal(pds)	467
194	coffee (cups)	402	244	coal(others)	468
195	coffee (powder)	403	245	l.p.g.	469
196	cold beverages/ bottled/canned	404	246	charcoal	470
197	fruit juice/shake	405	247	other oil used for lightening	471
198	coconut (green)	406	248	candle	472
199	other beverages (bournvita,horlicks,chocolate	407	249	methylated spirit	473
200	beverages (192-199)	409	250	gobar gas	474
CASH PURHAGE OF FOOD,PAN,TOBACCO,INTOXICANTS A			251	other fuel and light	475
201	biscuits and confectionery	410	252	fuel and light (total) (236-251)	479
202	salted food	411	CLOTHING		
203	prepared sweets	412	253	dhoti	480
204	coocked meals	413	254	sari	481
205	cake pastry	414	255	cloth for shirt, pajama etc	482
206	pickles	415	256	cloth for coat, trousers	483
207	sauce	416	257	chadar, dupatta	484
208	jam/jelly	417	258	lungi	485
209	other processed food	418	259	gamchha, towelE	486
210	processed food	419	260	hosiery articles	487
211	beverages all (192-199)	429	261	ready made garments	488
212	pan leaf	430	262	head ware	489
213	pan leaf(prepared)	431	263	knitted garmentsE	490
214	supari	432	264	bed sheetsE	491
215	lime	433	265	rug blanket	492
216		434	266	pillowE	493
217	other ingradient for pan	435	267	mosquito net	494
218	pan (s.t)(212-217)	439	268	matsE	495
219	bidi	440	269		496
220	sigeratte	441	270		497
221	leaf tobacco	442	271	clothing total (253-270)	509
222	snuff	443	FOOT WARE		
223	hookah tobacco	444	272	leather boots	510
224	churut	445	273	leather sandles	511
225	zarda,surtiE	446	274	others leather foot wear	512
226	other tobacco products	447	275	other rubber footwear	513
227	tobacco(219-226)	449	276	other foot wear (pair)	518
228	ganja	450	277	foot wear total (272-276)	519
229	toddy	451		EXPENDITURE ON MISC. GOODS & SERVICES,R	
230	country liquor	452	277	cinema, theatre	520
231	opium,bhangE	453	278	mela, fare	521
232	beer	454	279	sports goods toys etc.	522
233	foreign liquor	455	280	club fees	523
234	other drugs and intoxicants	456	281	goods for recreationE	524
235	intoxicants (228-234)	459	282	photography	525
FUEL AND LIGHT			283	other amusement	526
236	coke	460	284	amusement total (253-259)	529
237	firewood and chips	461	285	spectacles	530

B3. (Continued).

238	electricity	462	286	torch light	531
239	dung cake	463	287	fountain pen	532
240	kerosene (pds)	464	288	lock	533
289	umbrella, raincoat	534	341	bullock cart fare	606
290	walking stick	535	342	house cart fare	607
291	lighter	536	343	hand operatedE	610
292	asanitary napkin	537	344	porter charges	611
293	other goods for personal care and effect	538	345	petrol	612
294	goods for personal care and effect (261-269)	539	346	diesel	613
295	toilet soap	540	347	lubricating oil	614
296	toothbrush,tooth paste	541	348	other imputed value of owned conveyence	615
297	powder, snow ,cream	542	349	conveyence total (335-348)	619
298	hair oil, lotion, shampoo	543	350	misc. consumer setvices total (334+349)	629
299	comb	544	351	house rent, garage rent	630
300	shaving bladesE	545	352	residential land rent	631
301	shaving cream	546	353	consumer rent (other)	632
302	other shaving requisites	547	354	rent total (351-353)	639
303	other toilet articles	548	355	consumer taxes	640
304	toilet articles (271-279)	549	356	water charges	641
305	electric bulb, tube, light	550	357	other tax etc.	642
306	electric batterries	551	358	consumer taxes total (355-357)	649
307	other non-durables electric goods	552	EDUCATIONAL EXPENSES		
308	earthenware	553	359	books, journal	650
309	glass ware	554	360	news paper	651
310	plastic goods	555	361	library charges	652
311	coir, rope etc.	556	362	stationary articles	653
312	washing soap	557	364	tution fees	654
313	washing soda	558	365	pvt. Tutor	655
314	other washing requisites	559	366	other educational expenses	658
315	agarbati	560	367	educational total (359-366)	659
316	flower	561	NON-INSTITUTIONAL MEDICAL EXP.		
317	insecticide and related goods	562	368	medicine	660
318	other petty articles	563	369	family planning	661
319	sundry articles (281-294)	569	370	x-ray, ecgE	662
320	misc. con. Goods. Total (260+270+280+295)	579	371	nurse/mid-wife	663
321	cosmetics/servents/cook	580	372	other medical exp	668
322	sweper	581	373	non-ins. Medical total (368-372)	669
323	barbar, beautician	582	INSTITUTIONAL MEDICAL EXP.		
324	washerman	583	374	medicine	670
325	tailor	584	375	family planning	671
326	priest	585	376	x-ray, ecgE	672
327	legal exp.	586	377	doctor charges	673
328	postage & stamps	587	378	nurse/mid-wife	674
329	telephone	588	379	hospital charges	675
330	repair charges	589	380	nursing home/polyclinic	676
331	misc. expenses.	590	381	other medical exp	678
332	pet animals	591	382	ins. Medical total (374-382)	679
333	other consumer services	592	383	medical exp total (373+382)	689
334	consumer services total (321-333)	599	PURCHASE AND CONS. OF CONSUMER DURABLE GOO		
335	air ways fare	600	384	bed stead	690
336	railway fare	601	385	almirah	691
337	bus fare	602	386	dressing table	692

B3. (Continued).

338	taxi, auto rickshawE	603
339	steamerE	604
340	rickshaw hand driven& cycle	605
390	stool bench	696
391	suitcase, trunk..etc	697
392	foam, rubber, cushion	670
393	carpet and other floor matting	671
394	painting, drawingsE	672
395	other furniture and fixtures	673
396	furniture and fixture	674
397	gramaphone and record player	710
398	radio	711
399	radio gram	712
400	VCR/VCP	713
401	camera and other photography equipment	714
402	tape recorder	715
403	harmonium	716
404	piano	717
405	musical audio/vedeo cassatte	720
406	other musical instruments	721
407	other goods for recreation	728
408	goods for recreation(total)	729
409	electric fan	750
410	stove	751
411	pressure cooker	752
412	non-sticking pan	753
413	geyser	754
414	sewing machine	755
415	washing machine	756
416	refrigerator	757
417	air conditioner	760
418	lantern lamp	761
419	other electrical heating appliances	762
420	theremos	763
421	ovens etc	768
422	cooking and households appliance total	769
423	bicycle	770
424	motor cycle	771
425	motor, jeep	772
426	tyres, tubes	773
427	perambulator	774
428	other transport	778
429	personal transport total	779
430	hearing aids	780
431	glass eyes, artificial limbs etc	781
432	surgical belts etc	782
433	clinical thermometer	783
434	other therapeutic appliances	788
435	therap..total	789
436	clock, watch	790
387	chair	693
388	coach sofa	694
389	table desk	695
437	type writer	791
438	electric hair drier	792
439	other personal goods	798
440	personal goods total	799
441	residential building and land	800
442	other durable	808
443	residential building and land total	809
444	durables goods total	819

APPENDIX C

C1. Sector-Wise Distribution of Private Final Consumption Expenditure among Lower, Middle and Higher Income Groups for the Year 1983-84 (% share)

Sl No	Sectors	lower	middle	higher
1	Coal and lignite	5	75	20
2	Crude petroleum, natural gas	0	0	0
3	Electricity	4.32	35.765	59.9
4	Agricultural crops	19.65	23.17	57.16
5	Animal husbandry	31.57	7.14	61.27
6	Forestry and logging	30.35	12.36	57.28
7	Fishing	30.35	12.36	57.28
8	Metallic and non-metallic minerals	0	0	0
9	Food and food products	6.23	36.78	56.98
10	Beverages	0.09	36.49	63.4
11	Tobacco products	1.33	57.47	41.19
12	Cotton textiles	4.66	54.28	41.04
13	Wool, Silk and Synthetic fibre textiles	4.66	41.04	54.28
14	Jute, hemp, mesta textiles	0	5.447	94.55
15	Miscellaneous textile products	4.18	8.696	87.11
16	Wood and wood products including furniture etc.	5.98	29.45	64.56
17	Paper, printing and publishing	12.69	51.26	36.03
18	Leather and leather products	0	20.95	79.04
19	Rubber products	4.98	34.95	60.05
20	Plastic products	4.98	31.65	63.35
21	Petroleum products	2.36	38.78	58.84
22	Coal tar products	0	0	0
23	Inorganic heavy chemicals	0	0	0
24	Organic heavy chemicals	0	0	0
25	Fertilizers	0	0	0
26	Paints, varnishes and lacquers	0	0	100
27	Other chemicals	2.04	23.76	74.19
28	Cement	0	0	0
29	Clay and non-metallic mineral products	8.98	39.96	51.05
30	Iron and steel	0	0	0
31	Non-ferrous basic metals	0	0	100
32	Miscellaneous metal products	3.76	29.45	66.77
33	Tractors and agricultural machinery	0	0	0
34	Industrial machinery(F and T)	0	0	0
35	Other Machinery	0	20	80

C1. (Continued)

	Sectors	lower	middle	higher
36	Electrical and Electronic appliance	0	14.3	85.69
37	Rail equipments	0	0	0
38	Other transport equipments	8.96	34.16	56.87
39	Miscellaneous manufacturing	0	21.32	78.67
40	Construction	0	0	0
41	Gas and Water	4.32	35.76	59.9
42	Railway transport services	0	46.29	53.703
43	Other transport services	5.32	31.45	63.22
44	Storage and warehousing	0	0	0
45	Communication	0	30.56	69.45
46	Medical and health	5	30	65
47	Misc. Services	6.78	36.67	56.53

C2. Sector-Wise Distribution of Private Final Consumption Expenditure among Lower, Middle and Higher Income Groups for the Year 1989-90 (% share)

Sl No	Sectors	lower	middle	higher
1	Coal and lignite	8.59	68.67	22.74
2	Crude petroleum, natural gas*	0	0	0
3	Electricity	6.53	32.56	60.91
4	Agricultural crops	23.94	32.18	43.88
5	Animal husbandry	24.7	21.92	53.38
6	Forestry and logging	24.09	24.53	51.38
7	Fishing	23.13	25.68	51.2
8	Metallic and non-metallic minerals	0	0	0
9	Food and food products	9.45	28.72	61.83
10	Beverages	10.48	35.72	53.8
11	Tobacco products	5.19	45.36	49.45
12	Cotton textiles	6.94	39.67	53.38
13	Wool, Silk and Synthetic fibre textiles	2.33	33.25	64.42
14	Jute, hemp, mesta textiles	2.52	11.34	86.15
15	Miscellaneous textile products	2.09	16.21	81.7
16	Wood and wood products including furniture etc.	6.05	28.33	65.62
17	Paper, printing and publishing	9.91	44.61	45.48
18	Leather and leather products	4.07	24.08	72.86
19	Rubber products	6.06	36.45	57.49
20	Plastic products	6.06	34.8	59.15
21	Petroleum products	8.57	37.79	53.64

C2. (Continued).

Sl No	Sectors	lower	middle	higher
22	Coal tar products	0	0	0
23	Inorganic heavy chemicals	0	0	0
24	Organic heavy chemicals	0	0	0
25	Fertilizers	0	0	0
26	Paints, varnishes and lacquers	0	0	0
27	Other chemicals	11.31	28.33	60.36
28	Cement	0	0	0
29	Clay and non-metallic mineral products	15.21	36.96	47.83
30	Iron and steel	0	0	0
31	Non-ferrous basic metals	0	0	50
32	Miscellaneous metal products	1.88	20.28	77.84
33	Tractors and agricultural machinery	0	0	0
34	Industrial machinery(F and T)	0	0	0
35	Other Machinery	6.12	27.28	66.6
36	Electrical and Electronic appliance	1.72	18.6	79.68
37	Rail equipments	0	0	0
38	Other transport equipments	11.15	33.92	54.93
39	Miscellaneous manufacturing	9.33	24.78	65.89
40	Construction	0	0	0
41	Gas and Water	6.18	34.16	59.63
42	Railway transport services	8.16	39.58	52.26
43	Other transport services	4.05	18.26	77.69
44	Storage and warehousing	0	0	0
45	Communication	0	28.01	72
46	Medical and health	10.37	31.24	58.39
47	Misc. Services	8.73	32.9	58.37

*data was not provided for Crude petroleum and natural gas

C3. Sector-Wise Distribution of Private Final Consumption Expenditure among Lower, Middle and Higher Income Groups for the Year 1993-94 (% share)

Sl No	Sectors	lower	middle	higher
1	Coal and lignite	12.17	62.33	25.48
2	Crude petroleum, natural gas	5.034	17.234	77.74
3	Electricity	8.739	29.345	61.92
4	Agricultural Crops	28.225	41.17	30.59
5	Animal husbandry	17.82	36.68	45.48
6	Forestry and logging	17.82	36.68	45.48

C3. (Continued).

	Sectors	lower	middle	higher
7	Fishing	15.897	38.987	45.116
8	Metallic and non-metallic minerals	0	0	0
9	Food and food products	12.675	20.654	66.68
10	Beverages	20.85	34.95	44.192
11	Tobacco products	9.057	33.23	57.706
12	Cotton textiles	9.21	25.067	65.71
13	Wool, Silk and Synthetic fibre textiles	0	25.44	74.55
14	Jute, hemp, mesta textiles	5.03	17.234	77.74
15	Other textiles products	0	23.71	76.28
16	Wood and wood products including furniture	6.12	27.21	66.66
17	Paper, printing and publishing	7.12	37.94	54.93
18	Leather and leather products	8.13	27.21	66.66
19	Rubber products	7.12	37.94	54.93
20	Plastic products	7.12	37.94	54.93
21	Petroleum products	14.76	36.78	48.44
22	Coal tar products	0	0	0
23	Inorganic heavy chemicals	0	0	0
24	Organic heavy chemicals	0	0	0
25	Fertilizers	0	0	0
26	Paints, varnishes and lacquers	0	0	0
27	Other chemicals	20.56	32.89	46.53
28	Cement	0	0	0
29	Clay and non-metallic mineral products	21.42	33.95	44.61
30	Iron and steel	0	0	0
31	Non-ferrous basic metals	0	0	0
32	Miscellaneous metal products	0	11.1	88.89
33	Tractors and agricultural implements	0	0	0
34	Industrial machinery(F and T)	0	0	0
35	Other machinery	12.23	34.56	53.19
36	Electrical and Electronic appliances	3.43	22.89	73.67
37	Rail equipments	0	0	0
38	Other transport equipments	13.34	33.67	52.98
39	Miscellaneous manufacturing industries	18.66	28.23	53.1
40	Construction	0	0	0
41	Gas and Water	8.03	32.56	59.35
42	Railway transport services	16.31	32.85	50.82
43	Other transport services	2.77	5.06	92.161
44	Storage and warehousing	0	0	0
45	Communication	0.05	32.84	67.11

C3. (Continued).

	Sectors	lower	middle	higher
46	Medical and Health	15.73	32.47	51.78
47	Misc. Services	10.67	29.12	60.2

C4. Sector-Wise Distribution of Private Final Consumption Expenditure among Lower, Middle and Higher Income Groups for the Year 1998-99 (% Share)

Sl No	Sectors	lower	middle	higher
1	Coal and lignite	11.17	68.69	20.14
2	Crude petroleum, natural gas	0	0	0
3	Electricity	6.88	38.07	55.05
4	Agricultural Crops	19.95	49.03	31.02
5	Animal husbandry	11.13	38.16	50.71
6	Forestry and logging	30.12	16.86	53.02
7	Fishing	16.79	34	49.21
8	Metallic and non-metallic minerals	0	0	0
9	Food and food products	12.92	45.33	41.75
10	Beverages	11.4	44.16	44.44
11	Tobacco products	16.87	43.19	39.94
12	Cotton textiles	12.38	40.31	47.31
13	Wool, Silk and Synthetic fibre textiles	5.27	36.05	58.68
14	Jute, hemp, mesta textiles	3.25	17.33	79.42
15	Other textiles products	5.79	21.11	73.1
16	Wood and wood products including furniture	8.49	21.78	69.72
17	Paper, printing and publishing	6.77	42.29	50.94
18	Leather and leather products	5.53	30.94	63.52
19	Rubber products	6.45	28.44	65.11
20	Plastic products	6.96	37.35	55.69
21	Petroleum products	4.4	40.22	55.38
22	Coal tar products	15	33	52
23	Inorganic heavy chemicals	0	0	0
24	Organic heavy chemicals	0	0	0
25	Fertilizers	0	0	0
26	Paints, varnishes and lacquers	2.75	29.15	68.1
27	Other chemicals	2.75	29.15	68.1
28	Cement	0	0	0
29	Clay and non-metallic mineral products	15	33	52
30	Iron and steel	0	0	0
31	Non-ferrous basic metals	0	0	0

C4. (Continued).

Sl No	Sectors	lower	middle	higher
32	Miscellaneous metal products	2.66	20.46	76.88
33	Tractors and agricultural implements	30.26	55.34	14.4
34	Industrial machinery(F and T)	0	0	0
35	Other machinery	7.75	23.91	68.34
36	Electrical and Electronic appliances	2.33	25.65	72.02
37	Rail equipments	0	0	0
38	Other transport equipments	7.05	35.26	57.7
39	Miscellaneous manufacturing industries	7.75	23.7	68.55
40	Construction	0	0	0
41	Gas and Water	3.13	30.84	66.03
42	Railway transport services	3.49	50.42	46.08
43	Other transport services	3.78	34.22	62
44	Storage and warehousing	0	0	0
45	Communication	2.51	31.24	66.25
46	Medical and Health	9.34	21.29	69.37
47	Misc. Services	6.23	30.65	63.12

C5. Sector-Wise Distribution of Private Final Consumption Expenditure among Lower, Middle and Higher Income Groups for the Year 2003-4 (% Share)

Sl No	Sectors	lower	middle	higher
1	Coal and lignite	16.67	54.37	28.96
2	Crude petroleum, natural gas	9.12	36.84	54.04
3	Electricity	10.34	35.52	54.14
4	Agricultural Crops	18.82	47.54	33.65
5	Animal husbandry	7.12	40.44	52.44
6	Forestry and logging	18.61	48.51	32.88
7	Fishing	9.01	42.56	48.43
8	Metallic and non-metallic minerals	0.00	0.00	0.00
9	Food and food products	11.60	43.59	44.81
10	Beverages	7.06	35.10	57.84
11	Tobacco products	13.58	48.23	38.19
12	Cotton textiles	12.42	44.03	43.55
13	Wool, Silk and Synthetic fibre textiles	10.36	41.25	48.40
14	Jute, hemp, mesta textiles	12.42	44.03	43.55
15	Other textiles products	10.36	41.25	48.40
16	Wood and wood products including furniture	6.97	30.60	62.42
17	Paper, printing and publishing	6.97	36.60	56.42
18	Leather and leather products	8.66	40.23	51.12

C5. (Continued).

Sl No	Sectors	lower	middle	higher
19	Rubber products	7.66	35.72	56.62
20	Plastic products	10.25	45.31	44.44
21	Petroleum products	7.11	36.43	56.46
22	Coal tar products	0.00	0.00	0.00
23	Inorganic heavy chemicals	0.00	0.00	0.00
24	Organic heavy chemicals	0.00	0.00	0.00
25	Fertilizers	0.00	0.00	0.00
26	Paints, varnishes and lacquers	4.57	35.84	59.59
27	Other chemicals	3.66	35.72	60.62
28	Cement	0.00	0.00	0.00
29	Clay and non-metallic mineral products	5.57	35.84	58.59
30	Iron and steel	0.00	0.00	0.00
31	Non-ferrous basic metals	0.00	0.00	0.00
32	Miscellaneous metal products	6.25	36.25	57.50
33	Tractors and agricultural implements	0.00	0.00	0.00
34	Industrial machinery(F and T)	0.00	0.00	0.00
35	Other machinery	3.98	35.30	60.71
36	Electrical and Electronic appliances	5.54	30.87	63.59
37	Rail equipments	0.00	0.00	0.00
38	Other transport equipments	8.09	35.17	56.74
39	Miscellaneous manufacturing industries	4.82	25.51	69.67
40	Construction	2.66	35.72	61.62
41	Gas and Water	4.96	32.33	63.71
42	Railway transport services	4.57	28.22	67.21
43	Other transport services	3.81	25.20	71.00
44	Storage and warehousing	0.00	0.00	0.00
45	Communication	2.81	21.82	75.37
46	Medical and Health	4.86	29.63	65.51
47	Misc. Services	2.27	22.81	74.92

APPENDIX D

D1. Total Intensity Of CO_2 During 1983-84 to 2003-4(mt of CO_2)

Sl No	Sectors	1983-84	1989-90	1993-94	1998-99	2003-4
1	Coal and lignite	0.06577	0.10138	0.10578	0.06016	0.057748
2	Crude petroleum, natural gas	0.04441	0.05874	0.05991	0.042705	0.126798
3	Electricity	12.53178	15.09915	15.28331	14.6624	16.62331
4	Agricultural crops	0.00004	0.00006	0.00006	6.59E-05	0.000101
5	Animal husbandry	0.00002	0.00003	0.00002	2.46E-05	3.34E-05
6	Forestry and logging	0.00003	0.00002	0.00002	2.84E-05	3.54E-05
7	Fishing	0.00002	0.00004	0.00005	5.9E-05	8.74E-05
8	Metallic and non-metallic minerals	0.00019	0.00019	0.00008	7.06E-05	8.43E-06
9	Food and food products	0.00008	0.00009	0.00009	0.000118	0.000119
10	Beverages	0.00015	0.00014	0.00011	0.000146	0.000149
11	Tobacco products	0.00004	0.00007	0.00006	8.27E-05	6.56E-05
12	Cotton textiles	0.00013	0.00015	0.00012	0.00016	0.000169
13	Wool, Silk and Synthetic fibre textiles	0.00012	0.00015	0.00013	8.39E-05	0.000111
14	Jute, hemp, mesta textiles	0.00013	0.00016	0.00014	0.000136	0.000112
15	Miscellaneous textile products	0.00008	0.00011	0.00010	0.000106	0.000143
16	Wood and wood products including furniture etc.	0.00003	0.00004	0.00005	0.0001	9.69E-05
17	Paper, printing and publishing	0.00017	0.00019	0.00021	0.000207	0.000186
18	Leather and leather products	0.00006	0.00011	0.00008	8.58E-05	7.42E-05
19	Rubber products	0.00010	0.00014	0.00015	0.00012	0.000229
20	Plastic products	0.00012	0.00014	0.00014	8.99E-05	0.000203
21	Petroleum products	0.00167	0.00117	0.00163	0.001729	0.001694
22	Coal tar products	0.00197	0.00129	0.00109	0.001261	0.00137
23	Inorganic heavy chemicals	0.00034	0.00032	0.00029	0.000303	0.00035
24	Organic heavy chemicals	0.00033	0.00030	0.00028	0.00035	0.000253
25	Fertilizers	0.00069	0.00051	0.00046	0.000491	0.00049
26	Paints, varnishes and lacquers	0.00016	0.00019	0.00022	0.000236	0.000234
27	Other chemicals	0.00014	0.00017	0.00019	0.000223	0.000186
28	Cement	0.00042	0.00051	0.00042	0.000372	0.000361
29	Clay and non-metallic mineral products	0.00034	0.00032	0.00025	0.000235	0.000229
30	Iron and steel	0.00033	0.00045	0.00036	0.000307	0.000496
31	Non-ferrous basic metals	0.00024	0.00031	0.00024	0.000216	0.000461
32	Miscellaneous metal products	0.00014	0.00025	0.00020	0.00019	0.000254
33	Tractors and agricultural machinery	0.00018	0.00015	0.00018	0.000182	0.000215
34	Industrial machinery(F and T)	0.00016	0.00019	0.00015	0.000154	0.000242
35	Other Machinery	0.00015	0.00018	0.00016	0.000152	0.000217
36	Electrical and Electronic appliance	0.00012	0.00013	0.00014	0.000119	0.000145

D1. (Continued).

Sl No	Sectors	1983-84	1989-90	1993-94	1998-99	2003-4
37	Rail equipments	0.00007	0.00018	0.00014	0.000196	0.000209
38	Other transport equipments	0.00014	0.00015	0.00014	0.000127	0.000158
39	Miscellaneous manufacturing	0.00014	0.00012	0.00014	0.000174	0.000129
40	Construction	0.00011	0.00016	0.00014	0.000172	0.000254
41	Gas and Water	0.00023	0.00053	0.00006	5.28E-05	0.000154
42	Railway transport services	0.00018	0.00014	0.00017	0.000163	0.000147
43	Other transport services	0.00020	0.00017	0.00024	0.000367	0.000485
44	Storage and warehousing	0.00030	0.00012	0.00009	0.000136	0.000232
45	Communication	0.00002	0.00004	0.00004	2.4E-05	2.18E-05
46	Medical and health	0.00007	0.00009	0.00008	9E-05	7.36E-05
47	Misc. Services	0.00004	0.00015	0.00004	5.25E-05	3.77E-05

D2: Total Intensity of SO2 During 1983-84 to 2003-4 (mt of SO_2)

Sl No	Sectors	1983-84	1989-90	1993-94	1998-99	2003-4
1	Coal and lignite	0.0003841	0.0006124	0.0007289	0.000391	0.00042
2	Crude petroleum, natural gas	0.0003728	0.0004209	0.0005210	0.00037	0.001067
3	Electricity	0.0479019	0.0732488	0.0882466	0.061833	0.100599
4	Agricultural crops	0.0000003	0.0000004	0.0000005	5.83E-07	9.35E-07
5	Animal husbandry	0.0000002	0.0000002	0.0000002	2.1E-07	3.1E-07
6	Forestry and logging	0.0000002	0.0000002	0.0000002	2.52E-07	3.35E-07
7	Fishing	0.0000002	0.0000004	0.0000005	5.59E-07	8.58E-07
8	Metallic and non-metallic minerals	0.0000015	0.0000014	0.0000007	5.63E-07	5.92E-08
9	Food and food products	0.0000006	0.0000006	0.0000007	9.39E-07	1.05E-06
10	Beverages	0.0000009	0.0000008	0.0000007	9.4E-07	1.27E-06
11	Tobacco products	0.0000003	0.0000005	0.0000005	6.18E-07	5.71E-07
12	Cotton textiles	0.0000008	0.0000009	0.0000009	1.01E-06	1.44E-06
13	Wool, Silk and Synthetic fibre textiles	0.0000007	0.0000009	0.0000009	5.57E-07	9.16E-07
14	Jute, hemp, mesta textiles	0.0000009	0.0000010	0.0000010	9.54E-07	8.15E-07
15	Miscellaneous textile products	0.0000005	0.0000007	0.0000007	7.39E-07	1.2E-06
16	Wood and wood products including furniture etc.	0.0000002	0.0000003	0.0000004	6.83E-07	7.04E-07
17	Paper, printing and publishing	0.0000009	0.0000010	0.0000011	1.04E-06	1.35E-06
18	Leather and leather products	0.0000004	0.0000008	0.0000006	6.36E-07	6.42E-07
19	Rubber products	0.0000007	0.0000008	0.0000011	8.52E-07	1.91E-06
20	Plastic products	0.0000009	0.0000009	0.0000011	6.68E-07	1.64E-06
21	Petroleum products	0.0000176	0.0000123	0.0000172	1.8E-05	1.75E-05
22	Coal tar products	0.0000069	0.0000050	0.0000067	6.97E-06	9.95E-06
23	Inorganic heavy chemicals	0.0000019	0.0000018	0.0000018	2.05E-06	2.79E-06
24	Organic heavy chemicals	0.0000030	0.0000018	0.0000020	2.77E-06	2.03E-06
25	Fertilizers	0.0000054	0.0000042	0.0000042	4.38E-06	4.66E-06

D2. (Continued).

Sl No	Sectors	1983-84	1989-90	1993-94	1998-99	2003-4
26	Paints, varnishes and lacquers	0.0000012	0.0000012	0.0000016	1.68E-06	1.92E-06
27	Other chemicals	0.0000010	0.0000012	0.0000014	1.55E-06	1.6E-06
28	Cement	0.0000017	0.0000021	0.0000019	1.6E-06	2.05E-06
29	Clay and non-metallic mineral products	0.0000020	0.0000020	0.0000016	1.6E-06	1.79E-06
30	Iron and steel	0.0000016	0.0000020	0.0000018	1.48E-06	2.45E-06
31	Non-ferrous basic metals	0.0000017	0.0000016	0.0000014	1.19E-06	2.24E-06
32	Miscellaneous metal products	0.0000007	0.0000013	0.0000012	1.04E-06	1.33E-06
33	Tractors and agricultural machinery	0.0000011	0.0000008	0.0000011	1.12E-06	1.25E-06
34	Industrial machinery(F and T)	0.0000009	0.0000011	0.0000011	1.03E-06	1.47E-06
35	Other Machinery	0.0000008	0.0000010	0.0000010	8.95E-07	1.3E-06
36	Electrical and Electronic appliance	0.0000008	0.0000008	0.0000009	7.63E-07	9.05E-07
37	Rail equipments	0.0000004	0.0000011	0.0000008	1.07E-06	1.27E-06
38	Other transport equipments	0.0000008	0.0000009	0.0000009	7.88E-07	1E-06
39	Miscellaneous manufacturing	0.0000009	0.0000007	0.0000010	1.09E-06	9.77E-07
40	Construction	0.0000006	0.0000008	0.0000009	1.04E-06	1.74E-06
41	Gas and Water	0.0000013	0.0000050	0.0000004	3.13E-07	1.09E-06
42	Railway transport services	0.0000012	0.0000009	0.0000009	9.85E-07	1.06E-06
43	Other transport services	0.0000020	0.0000016	0.0000023	3.47E-06	4.86E-06
44	Storage and warehousing	0.0000019	0.0000007	0.0000006	7.67E-07	1.58E-06
45	Communication	0.0000001	0.0000002	0.0000003	1.59E-07	1.56E-07
46	Medical and health	0.0000005	0.0000006	0.0000006	6.64E-07	6.45E-07
47	Misc. Services	0.0000002	0.0000010	0.0000003	4.01E-07	3.18E-07

D3: Total Intensity of NO_x During 1983-84 To 2003-4(mt of NO_x)

Sl No	Sectors	1983-84	1989-90	1993-94	1998-99	2003-4
1	Coal and lignite	0.0012378	0.0018328	0.0015825	0.000964	0.000762
2	Crude petroleum, natural gas	0.0004188	0.0008199	0.0004985	0.000335	0.001132
3	Electricity	0.3287823	0.3395321	0.2913616	0.361554	0.295818
4	Agricultural crops	0.0000005	0.0000007	0.0000006	4.7E-07	5.99E-07
5	Animal husbandry	0.0000003	0.0000004	0.0000002	2.03E-07	1.91E-07
6	Forestry and logging	0.0000002	0.0000002	0.0000002	2.02E-07	1.74E-07
7	Fishing	0.0000001	0.0000003	0.0000002	2.81E-07	3.16E-07
8	Metallic and non-metallic minerals	0.0000022	0.0000025	0.0000007	7.39E-07	1.19E-07
9	Food and food products	0.0000012	0.0000014	0.0000010	1.23E-06	8.79E-07
10	Beverages	0.0000026	0.0000028	0.0000019	2.37E-06	1.28E-06
11	Tobacco products	0.0000006	0.0000012	0.0000008	1.02E-06	5.15E-07
12	Cotton textiles	0.0000023	0.0000027	0.0000016	2.65E-06	1.43E-06
13	Wool, Silk and Synthetic fibre textiles	0.0000023	0.0000027	0.0000020	1.3E-06	1.05E-06
14	Jute, hemp, mesta textiles	0.0000020	0.0000030	0.0000019	1.92E-06	1.49E-06

D3. (Continued).

Sl No	Sectors	1983-84	1989-90	1993-94	1998-99	2003-4
15	Miscellaneous textile products	0.0000015	0.0000019	0.0000012	1.53E-06	1.28E-06
16	Wood and wood products including furniture etc.	0.0000004	0.0000005	0.0000007	1.48E-06	1.28E-06
17	Paper, printing and publishing	0.0000036	0.0000042	0.0000044	4.48E-06	2.48E-06
18	Leather and leather products	0.0000008	0.0000018	0.0000010	1.08E-06	5.94E-07
19	Rubber products	0.0000014	0.0000025	0.0000021	1.63E-06	2.11E-06
20	Plastic products	0.0000017	0.0000023	0.0000018	1.13E-06	2.04E-06
21	Petroleum products	0.0000024	0.0000019	0.0000024	2.23E-06	2.78E-06
22	Coal tar products	0.0000538	0.0000336	0.0000190	2.49E-05	1.82E-05
23	Inorganic heavy chemicals	0.0000067	0.0000063	0.0000052	4.56E-06	3.7E-06
24	Organic heavy chemicals	0.0000020	0.0000054	0.0000038	3.73E-06	2.64E-06
25	Fertilizers	0.0000079	0.0000050	0.0000034	3.35E-06	2.32E-06
26	Paints, varnishes and lacquers	0.0000020	0.0000035	0.0000030	3.22E-06	2.29E-06
27	Other chemicals	0.0000019	0.0000027	0.0000026	3.18E-06	1.51E-06
28	Cement	0.0000106	0.0000130	0.0000102	9.06E-06	6.93E-06
29	Clay and non-metallic mineral products	0.0000064	0.0000057	0.0000043	3.52E-06	2.56E-06
30	Iron and steel	0.0000073	0.0000106	0.0000079	6.88E-06	1.09E-05
31	Non-ferrous basic metals	0.0000037	0.0000063	0.0000044	4.27E-06	1.03E-05
32	Miscellaneous metal products	0.0000027	0.0000053	0.0000039	3.79E-06	5.31E-06
33	Tractors and agricultural machinery	0.0000035	0.0000031	0.0000030	3.15E-06	4.03E-06
34	Industrial machinery(F and T)	0.0000030	0.0000037	0.0000023	2.37E-06	4.31E-06
35	Other Machinery	0.0000028	0.0000036	0.0000028	2.8E-06	3.91E-06
36	Electrical and Electronic appliance	0.0000020	0.0000024	0.0000023	1.96E-06	2.47E-06
37	Rail equipments	0.0000012	0.0000034	0.0000027	3.91E-06	3.73E-06
38	Other transport equipments	0.0000024	0.0000027	0.0000023	2.19E-06	2.65E-06
39	Miscellaneous manufacturing	0.0000026	0.0000021	0.0000021	2.93E-06	1.55E-06
40	Construction	0.0000023	0.0000034	0.0000026	3.06E-06	3.75E-06
41	Gas and Water	0.0000046	0.0000032	0.0000010	9.62E-07	2.15E-06
42	Railway transport services	0.0000029	0.0000025	0.0000036	2.91E-06	1.97E-06
43	Other transport services	0.0000007	0.0000009	0.0000015	1.78E-06	1.39E-06
44	Storage and warehousing	0.0000051	0.0000024	0.0000015	2.63E-06	3.49E-06
45	Communication	0.0000003	0.0000006	0.0000006	3.74E-07	2.95E-07
46	Medical and health	0.0000010	0.0000014	0.0000010	1.14E-06	5.58E-07
47	Misc. Services	0.0000006	0.0000025	0.0000005	6.16E-07	3.34E-07

APPENDIX E

QUESTIONNAIRE FOR HOUSEHOLD SURVEY

Block A

General features of the selected location
Date of entry: ----------------------------- Area code: ---------------------------
Air pollution monitoring station: --
Enumerator's Name: --------------------- Address: -----------------------------

Socio-Economic Characteristics
1. Name of the area
Slum/non slum
2. Name of the town
3. Name of the corporation/municipality
4. Name of the police stations
5. Total number of households
6. Total number of population male female

7. Total number of educational institutions
Primary
Secondary
Higher secondary
College
Others, if any specify

8. Medical services
Number of doctors--- allopathic---- homeopathy----- others
Number of hospitals---- government ----private
Number of nursing home
Number of pathology and diagnostic Centre
Number of medical shops

9. Mortality rate in the area
Total death rate
Age specific death rate male female
Under 5 years
5-14
15-24
25-44
45-64
65+
All ages

10. Total number of industries---
large medium small

11. Mode of transport plying in the area
Bus /car/auto rickshaw/motor bike /heavy vehicles

Block B

Household
1. a) Name of the respondent: ---------------------------------------
b) Address: --
c) Telephone no: ---

2. a) For how long you are staying in this house?
3. Occupation of the head of the household and respondent: -
a) a. Office, b. Factory,
b) Self-employed: a. Office, b. Factory
c) Unemployed
d) Others (Please mention).

4. Total family members male female

5. Age distribution
Age of the respondent
Male female
Under 5 years
5-14
15-24
25-44
45-64
65+
All ages

6. A) Income of the household (monthly)
Income of the respondent (monthly)

B) Total expenditure of the household (monthly)

7. a) Distance of the house from the main road

b)Pattern of the house
Kacha
Pucca
Area of the house (sq ft)
Number of rooms

Area of the rooms (sq ft)
Ventilation yes no
Kitchen --- separate not separate

8. Energy consumption pattern

a) Cooking (monthly)
Type of energy used quantity value
Coal
Dung cake
Kerosene
Electricity
LPG
Solar cooker
Wood, paper etc
Others, if any specify

b) Lighting
Candle
Diesel generator
Electricity
Kerosene
Others, if any specify

Total energy consumption expenditure monthly
b) Number of cooking in a day once twice or more
c)Total cooking time in a day (hrs)

9. Health related data
9.1. General Health question
i) Are you aware that air pollution causes illness? Yes/ No
ii) Kindly mark the diseases you attribute to air pollution in the list of diseases given below
a) Headache
b) Eye /nose/throat irritation
c) Runny nose/Cold
d) Flu / Fever
e) Skin infection/Rash
f) Asthma attacks
g) Shortness of breadth
h) Respiration allergy to dust andpollen
i) Dry scratchy throat
j) Chest pain
k) Cough with phlegm
l) Dry cough
m) Bronchitis

n) Drowsiness
o) Pneumonia
p) Heart Disease
q) Cancer

9.2. Health History of the Household

(i) Have you or any one of your family members suffered from some of these diseases during

Last week Last month Last six month
Yes/No Yes /No Yes/No

If yes mention the name of the diseases:
Is it chronic? Yes /No

(ii) How many days were you sick due to the above mentioned diseases during?
Last week Last month Last six months
-------------- ----------------- --------------------

(iii) How many days other family members were sick?

(Suppose if two adults are sick for two days in the last week, the total number of sick days for adults are four days. Similarly for children).

Last week Last month Last six months
A C A C A C

(iv) Did you and /or any of your family members suffer from the respiratory diseases during?

Last week Last month Last six months
Yes/No Yes/No Yes/No

9.3. Health Stock: Status of health prior to August 2004

i) Do you smoke? Yes/No

ii Do you drink alcohol? Yes/No

Male female
iii) Hours of stay at home
Age distribution
Under 5 years
5-14
15-24
25-44
45-64
65+
All ages

9.4. Outdoor Pollution Averting Activities (you/or any of your family member)

i) Do you take any of these measures to avoid exposure to pollution?

a) Staying indoors Yes/No

b) Using mask while on road Yes/No

c) Avoiding busy roads and busy timings for the local travel Yes/No

If Yes, how many extra km you travel every day to avoid busy polluted areas?............

Any other measures that you consciously take up to avoid exposure to air pollution, kindly mention ....

ii) Do you think that outdoor pollution has affected your recreation? Yes/No

iii) Do you regularly exercise? Yes/No

iv) Do you go for morning or evening walk? Yes/No

9.5. Indoor Pollution

i) Do you use air conditioner at home? Yes/No

ii) Do you use heater during the winter? Yes/No

iii) Do you have exhaust fan or chimney in your kitchen? Yes/No

9.6 Monthly Observations (respondent/disease person)

a) Number of days suffering

b) Visit to doctors --- number of times

c) Hospital visits---

i) Number of times

ii) Number of days in hospital

iii) Emergency room visit

d) Nursing home visit---

i) Number of times

ii) Number of days in nursing home

iii) Emergency room visit

e) Restricted activity days

i) Days spent in bed

ii) Days missed from work

iii) Number of days when normal activities are restricted due to illness even if medical attention is not required

f) Total medical expenses

i) Expenses on doctor

ii) Expenses on medicine

iii) Expenses on special medicinal food

iv) Expenses on pathological test

g) Number of working days lost

i) Education

ii) Office
iii) Household work
iv) Loss of income due to absence from work owing to disease

h) Special information
i) Number of deaths in the family, if any (last 1 year)
ii) Causes of death
iii) Age at death
Employed
Unemployed

9.7. How many members in your family have health insurance?

9.8. Number of sick /medical leaves you and other adult employed members have in a year?

REFERENCES

ADB (2006) Country synthesis report on urban air quality management, India, Discussion draft, December, published by ADB, Philippine

Agarwal, A., Naryan, S. and Sen, S. (1999) Citizens' Fifth Report part I and part II, State of India's environment, Centre for Science and Environment, New Delhi.

Agarwal, A. (1999) Engines of the Devil: Why dieselisation of private automobile fleet should be banned-The case study of Delhi, Centre for Science and Environment, New Delhi. Available at http://www. cseindia.org/campaign/apc/pdf/dieselmo.pdf.

Alberini, A et al., (2000) Valuing health effects of air pollution in developing countries: The case of Taiwan, *Journal of Environment Economics and Management*, 34: 107-126

American Thoracic Society (ATS) (1996) A committee of the environmental and occupational health assembly of the American Thoracic Society, State of the art review: health effects of outdoor air pollution, *American Review of Critical Care Medicine*, 153(1):33-50, and 153(2):477-998.

Anglo, E. (2004) Modeling the dispersion of particulates in Metro Manila, presented at Sweeping the dust off Manila's air: A science and policy dialogue 21st April 2004, www.observatory.ph/aq/presentations/07_Anglo_modeling/index.html

Anon (1997) Tiny killers, in Down to Earth, Society for environmental communications, New Delhi, Nov 15 p 29.

Awasthi, S., Glick, H. A., Fletcher, R. H. and Ahmed, N., (1996) Ambient air pollution and respiratory symptoms complex in preschool children, *Indian Journal of Medical Research*, 104: 257–262.

BBC News (2004) Congestion charging a success" BBC NEWS, 17th Feb. 2004, British Broadcasting Corporation, London, http://news.bbc.co.uk/1/hi/england/london/3494015.stm

BBC (2006) Case studies of other charging zones British Broadcasting Corporation, London, www.bbc.co.uk/london/congestion/cities.html

Bhaumik S(2007)Air pollution suffocates Calcutta, BBC News, Calcutta, http://news.bbc.co.uk/1/hi/world/south_asia/6614561.stm

BJEPB(2005) Annual environmental bulletin of Beijing, Beijing Environmental Protection Bureau, http://www.bjepb.gov.cn

Bobak M. (2000) Outdoor air pollution, low birth weight, and prematurity, *Environmental health perspectives*, 108: 173–176

Borja-Aburto, V.H. et al., (1997) Ozone, suspended particulates, and daily mortality in Mexico City. *American Journal of Epidemiology*, 145: 258–268.

Bossier E. and. Rous, R.D (1992) Economic effects of carbon tax: application in Belgium with the multi-sectoral model Hermes, *Energy Economics,* 14 :33–41.

Brandon, C. and Homman, K. (1991-92) Valuing environmental costs in India: the economy wide impacts of environmental degradation, World Bank, mimeo.

Brandon, C. (1997)Economic valuation of air and water pollution in Bangladesh.Workshop Discussion Draft, The World Bank , Washington, D. C.

Breuil, J.M. (1992) Input–Output analysis and pollutant emissions in France, *Energy Journal,* 13(3): 173–184.

Brimblecombe, P(1987) *The big smoke*, London , Methuen, 185pp

Burnett, R.T., Cakmak, S., Brook, J.R. (1998) The effect of the urban ambient air pollution mix on daily mortality rates in 11 Canadian cities. *Can. J. Public Health*. 89:152-6.

Calcutta News(2005) http://silkboard.wordpress.com/2007/02/22/mumbai-best-kolkata-worst/

CEN (2003) Health impacts of Kathmandu's air pollution, Factsheet 6, Clean Energy Nepal, Kathmandu, Nepal, www.keva.org.np/publicationlKEVA·s%20Publication/CEN%20Fact%20Sheet%206%20-%20Health_Eng.pdf.

Census of India (2001) Registrar General and Census Commissioner, New Delhi , India Center for Transportation Studies, Working paper series, Working Paper # CTS2004D

CSO (1990) Input-Output transaction table for 1983-84, Central Statistical Organisation Government of India, New Delhi

CSO (1997) Input-Output transaction table for 1989-90 of India, Central Statistical Organisation, Ministry of Statistics & Programme Implementation, Government of India, New Delhi.

CSO (2000) Input-Output transaction table for 1993-94 of India, Central Statistical Organisation, Ministry of Statistics & Programme Implementation, Government of India, New Delhi.

CSO (2005) Input-Output transaction table for 1998-99 of India, Central Statistical Organisation, Ministry of Statistics & Programme Implementation, Government of India, New Delhi.

CSO (2008) Input-Output transaction table for 2003-4 of India, Central Statistical Organisation , Ministry of Statistics & Programme Implementation, Government of India, New Delhi.

CSO(1999) New Series on National Accounts Statistics (Base year 1993-94), Central Statistical Organization, Ministry of Statistics & Programme Implementation, Government of India,.New Delhi, India

CSO(2005) New Series on National Accounts Statistics (Base year 1993-94), Central Statistical Organization, Ministry of Statistics & Programme Implementation, Government of India, New Delhi, India

CSO(2006), Energy Statistics Report, Central Statistical Organisation , Ministry of Statistics & Programme Implementation, Government of India, New Delhi, India

Chakraborti, D (2003) 'Kolkata city : An urban air pollution perspectives' In An Earthscan Readers In World Transport Policy And Practice, edited by Whitelegg J and Haq G London, Earthscan

Chameides, W., Kasabhatla, P., Yienger, J. and Levy, H.(1994) Regional ozone pollution and food production, *Science* 264 : 74–77

Chandrasiri S.(1999) Controlling automotive air pollution: the case of Colombo City, Economy and Environment Programme for Southeast Asia, Research Report Series, Colombo, Sri Lanka

Chang C.C.S.S., Ho , S.C. and Yang, C.Y.(2005) Air pollution and hospital admissions for cardiovascular disease in Taipei, Taiwan, *Environmental Research*, 98(1): 114-119

Chang, G., Pan, X., Xie, X. and Gao, Y. (2003) Time-series analysis on the relationship between air pollurion and the daily mortality in Beijing' (in Chinese), *Wei Sheng Yan Jiu*, 32:565-568.

Chang, Y.F. and Lin, S.J., (1998) Structural decomposition of industrial CO_2 emission in Taiwan: an input–output approach. *Energy Policy*, 26 (1), 5– 12

Chabra, S.K (1996) Prevalence of bronchial asthma in school children of Delhi, Vallabhbhai Patel Chest Institute, New Delhi.

Chabra, S. K., Chabra, P., Rajpal, S. and Gupta, R. K. (2001) Ambient air pollution and chronic respiratory morbidity in Delhi, *Archives of Environmental Health*, 56:58-64.

Chitkara, S. (1997) Air pollution in India – An exploratory study, National Compendium Statistics, New Delhi, India.

Civic Exchange (2004) Air pollution and air quality management issues in the Hong Kong and pearl river delta, civic exchange white paper, November,2004 Hong Kong, www.civic-exchange.org/publications/2004/airpollutionwhitepaper.pdf

Cohen A. J. et al., (2005) The global burden of disease due to outdoor air pollution, *Journal of Toxicology and Environmental Health*, Part A 68: 1301-07

Common, M.S. and Salma, U (1992) Accounting for changes in Australian carbon dioxide emissions. *Energy Economics*, 14 (3):217– 225.

Conceiao, G M S, Miraglia, S G E K, Kishi, H S, Saldiva, P N H and Singer, M (2001) Air pollution and child mortality: a time series study in Sao Paulo, Brazil, *Environmental Health Perspectives*, 109(3):347-350.

CORE (2008) www.indiacore.com

CPCB (2000) National ambient air quality status, Central Pollution Control Board, Ministry of Environment and Forest, New Delhi

CPCB(2001) Air quality in Delhi (1989–2000), Central Pollution Control Board, Ministry of Environment and Forest, New Delhi

CPCB(2004) Air quality of Delhi; 2004, Central Pollution Control Board, Ministry of Environment and Forest, New Delhi

Cropper et al (1997) The health effects of air pollution in Delhi, India, *PRD Working paper # 1860*, World Bank.

CSE (1997) The slow murder continues, the report published by Centre for Science and Environment, New Delhi

CSE (2006) The leapfrog factor: cleaning air in Asian cities, Centre for Science and Environment, New Delhi, India,

CSE (2006a) CSE draft dossier: health and environment, *The environment and poverty burden of disease,* Centre for Science and Environment, New Delhi

CSLF (2006) An energy summary of India, May, Carbon Sequestration Leadership Forum, http://www.cslforum.org/india.htm

Delmeek, J, Selevan, S G, Benes, 1, Solansky, I and Sram, R(1999) Fetal growth and maternal exposure to particulate matter during pregnancy, *Environmental Health perspectives*, 107 (6):475-480.

DENR (2005) National air quality status report (2003-2004), Department of Environment and Natural Resources, Manila (draft).

DEP (2003) Environmental monitoring and analysis, Taipei City, Department of Environmental Protection, http://l63.29.36.5l/dep/index.jsp?recordid=2208.

DEP (2004) Air quality of Taipei City in 2003, Taipei City, Department of Environmental Protection, http://l63.29.36.51/dep/index.jsp?recordid=2232.

Dincer, I (2000) Renewable energy and sustainable development: a crucial review, *Renewable and Sustainable Energy Reviews* 4 : 157-175

Dockery DW, Pope CA III, Xu X, Spengler JD, Ware JH, Fay ME, Ferris BG Jr, Speizer FE.(1993) An association between air pollution and mortality in six U.S. cities, *New England Journal of Medicine*, Dec 9,329(24):1753-9

Dockery DW, Pope, CAIII. (1994) Acute respiratory effects of particulate air pollution, *Ann Rev Pub Hlth* 15:107-13.

DOE/EIA (2006) Department of Energy, Official Energy Statistics from the US Govt. Energy Information Administration

Earth Trends (2006) Monthly update: Sustainable transport in the developing world, December, World resource institute, http://earthtrends.wri.org/updates/node/135

Economic Survey of Delhi: 2003–04, Government of National Capital Territory of Delhi, 6th Level, B-Wing, Delhi Secretariat, I. P. Estate, New Delhi, India. http://delhiplanning.nic.in/.

EIA(2007), Energy Information Administration, International Energy Annual-long term International Energy Statistics, Official Energy Statistics from the US Government

EIA(2007a), Energy information administration, International Energy Annual, Annual Energy outlook, Official Energy Statistics from the US Government, December release

Environmental Pollution in India(2008) http://www.gits4u.com/envo/envo4.htm

EPD (2004A) Environmental Protection Department, www.epd.gov.hk/epd /English/environmentinhk/air/data/emission_inve.html

EPD(2004b) Air quality in Hong Kong 2003, Environmental Protection Department, Hong Kong, www.epd-asg.gov.hk /english/report/files/aqr03e.pdf

Eskeland, G.S. and Deverajan, S. (1996) Taxing bads by taxing goods: pollution control with presumptive charges: directions in development series. Washington DC, World Bank

Eskeland, Gunnar A., Ostro, Bart D., Feyzioglu, Tarhan and Sanchez, Jose Miguel (1998), Air pollution and health effects: a study of respiratory illness among children in Santiago, Chile. *World Bank Policy Research Working Paper No. 1932*, March.

Europa(2005), The Kyoto Protocol, Press release rapid, MEMO/05/49, Brussels, 16 February

Fadel M., and Massoud, M(2000). Particulate Matter in urban areas: health-based economic assessments, *The Science of the Total Environment*, 257, : 133-146

Fairly D. (1990) The relationship of daily mortality to suspended particulates in Santa Clara County in 1980-1986, *Environmental Health Perspectives* .89:88-100.

FAO (2006) Rapid growth of selected Asian economies: Lessons and implications for agriculture and food security China and India, Regional Office for Asia and the Pacific Food and Agriculture Organization of the United Nations, Bangkok,

Garg, A.,P. R. Shukla, S. Bhattacharya, and V. K. Dadhwal (2001) Sub-region (district) and sector level SO_2 and NO_x emissions for India: assessment of inventories and mitigation flexibility ,*Atmospheric Environment* ,35(4): 703-713

Gauderman, W. J. et al. (2000) Association between air pollution and lung function growth in Southern California children, *American Journal of Respiratory and Critical Care Medicine*,162: 1383–1390 .

Gay, S., and Proops, J.L.R., (1993) Carbon dioxide production by UK economy: an input–output assessment, *Applied Energy*, 44 (2):113– 130.

GEO (2000) Alternative Policy Study: Reducing air pollution in Asia and the Pacific, Global Environment Outlook, United Nation Environment Programme

Ghose M. K (2002) Controlling of motor vehicle emissions for a sustainable city, *TERI Information Digest on Energy and Environme*nt 1(2): 273.282

Goldemberg, J. et al., (1993). 'The Brazilian fuel alcohol program' In *Renewable Energy: Sources for Fuels and Electricity*.edited by Johansson, T.B. Kelly, H., Kelly, H., Reddy, A.K.N., Williams, R., Island Press, Washington, DC.

GOI (1992) Eighth Five Year Plan 1992-97, The Planning Commission, Government of India of India, New Delhi.

Gouveia, N. and Fletcher T(2000) Time series analysis of air pollution and mortality: effects by cause, age and socioeconomic status, *Journal of epidemiology and community health*, 54: 750–755.

GTZ Energy Information, (2005) Fuel prices in ASEAN countries, http://www.international-fuel-prices.com/downloads/ASEANFuelPrices2005.pdf

Guo, Y, Lo, Lin, Y. C., Sung, F. C., Huang, S. L., Ko, Y. C., Lai, J. S., Su, H. J., Shaw, C. K., Lin, R. S. and Dockery, D. W. (1999) Climate, traffic-related air pollutants, and asthma prevalence in middle-school children in Taiwan, *Environmental Health Perspectives*, 107(12):1001-1006.

Gupta, U (2006) Valuation of urban air pollution: a case study of Kanpur City in India, SANDEE Working Paper No. 17-06 http://www.sandeeonline.org/publications/working_papers/wp17.pdf

Gurjar, B. R., van Aardenne, J. A., Lelieveld, J. and Mohan, M.(2004), Emission estimates and trends (1990–2000) for megacity Delhi and implications, *Atmospheric Environment* 38, 5663– 5681.

Gurjar, B. R.; A. Singh Nagpure and T. P. Singh; H. Hanson (2008) "Air quality in megacities." In: Encyclopedia of Earth. Eds. Cutler J. Cleveland (Washington, D.C.: Environmental Information Coalition, National Council for Science and the Environment).

Hall, D.O. et al. (1993) "Biomass for energy: supply prospects" In *Renewable Energy: Sources for Fuels and Electricity*.edited by Johansson, T.B. Kelly, H., Kelly, H., Reddy, A.K.N., Williams, R., Island Press, Washington, DC :593-651.

Han Xiaoli and T.K. Lakshmanan, (1994) Structural changes and energy consumption in the Japanese economy 1975-85: an input-output analysis, *The Energy Journal*, 15(3): 16-188.

Hann, M.D. (2001) A structural decomposition analysis of pollution in the Netherlands, *Economic Systems Research*, 13(2):181–196.

Hansen, A., Selte, C., Harald, K. (2000). Air pollution and sick-leaves: a case study using air pollution data from Oslo. *Environmental and Resource-Economics* 16(1):31-50.

Hao. J, J. Hu, L. Fu (2006), Controlling vehicular emissions in Beijing during the last decade, *Transportation Research Part A* 40 (2006) 639–651

Hasselblad, V., D.J. Kotchmar and D.M. Eddy (1992), Synthesis of environmental evidence: nitrogen dioxide epidemiology studies, *J. Air Waste Manage. Assoc*. 42 662–671

Hayami, H., et al., (1993) Estimation of air pollutions and evaluating CO_2 emissions from production activities: using Japan's 1985 input–output tables, *Journal of Applied Input–Output Analysis* 1(2), 29– 44.

HEI (2004) Helath effects of outdoor air pollution in developing countries of Asia:A literature review, HEI special report 15, Boston, MA, Health effects institute

Hong, Y.-C. et al., (1999) PM_{10} exposure, gaseous pollutants, and daily mortality in Inchon, South Korea. *Environmental health perspectives*, 107: 873–878.

Hooda, N and Rawat, V.R.S.(2006) Role of bio-energy plantations for carbon-dioxide mitigation with special reference to India, *Mitigation and Adaptation Strategies for Global Change* 11: 445–467

Hwang, J. So, Chen, Y. J., Wang, 1. D., Lai, Y. M., Yang, C. Y. and Chan, C. C. (2000) Subject-domain approach to the study of air pollution effects on schoolchildren's illness absence, *American Journal of Epidemiology*, 152(1):67-74

Ilabaca, M, Olaeta, 1, Campos, E, Villaire, J, Tellez-Rojo, M M and Romieu, I (1999), Association between levels of fine particulate and emergency visits for pneumonia and other respiratory illnesses among children in Santiago, Chile, *Journal of Air Waste Management Association*, 48: 174-183

IPCC(2007)Summary for Policymakers of the Synthesis Report of the IPCC Fourth Assessment Report, AR4 Draft Copy, November 2007

Islam, S. (2003) Dhaka gets AQI, The Daily Star, 4:55, 20 July 2003

Island The (2003) The deadly danger of urban air pollution, the island, Sunday 18th May www.southasianmedia.net/

Jinsart W., Tamure K., Loetkamonwit S., Thepanondh, S., Karita, K. and Yano, E. (2002) Roadside particulate air pollution in Bangkok, *Journal of Air Waster Management Association*, 52: 1102-1110

Kamat, S. R. (2000) Mumbai studies of urban air pollution and health resulting synergistic effects of SPM, S02 and NO, International Conference on Environmental and Occupational Respiratory Diseases, Indian Toxicological Research Centre, Lucknow, India.

Kan, H. D., Chen, B. H., Fu, C. W., Yu, S. Z. and Mu, L. N. (2005) Relationship between Ambient air pollution and daily mortality of SARS in Beijing, *Biomedical and Environmental Sciences*, 18:1-4

Karimzadegan H, M Rahmatian , DD Farhud , M Yunesian(2008) Economic Valuation of Air Pollution Health Impacts in the Tehran Area, Iran, *Iranian J Publ Health*, 37(1):20-30

KFW (2004) 'Nepal: Rehabilitation of Chobhar Cement Factory', ('Nepal: Rehabilitierung der Zementfabrik Chobhar Abschlusskontroll- und Schlussprufung Projekt Rehabilitierung der Zementfabrik Chobhar', in German), Kreditanstalt fuer Wiederaufbau - Enwicklungsbank, www.kfw-entwicklungsbank.deIDE_Home/ Ev al uieru ng/Weitere_Informationen/Schl uss prf90/PD F- Dokumente/nepal_ chobhar.pdf.

Kjellén M (2001) Health and Environment, Health Division Document 2001:2, SIDA publication

Klemm RJ, Mason RM Jr, Heilig CM, Neas LM, Dockery DW(2000) Is daily mortality associated specifically with fine particles? Data reconstruction and replication of analyses, *J Air Waste Manag Assoc*,50(7):1215–1222

Krupnick, Alan J. (1997) 'Urban Air Pollution in Developing Countries: Problems and Policies,' In *The environment and emerging development issues*, edited by Partha Dasgupta and Karl-Goran Maler, vol. 2. United Nations University/World Institute for Development Economics Research (UNU/WIDER) Studies in Development Economics. Oxford and New York: Oxford University Press, Clarendon Press, 1997.

Kumar R; Sharma M; Srivastva A ; Thakur J ; Jindal S; Parwana HK(2004) Association of outdoor air pollution with chronic respiratory morbidity in an industrial town in northern India, Archives of environmental health 59(9), :471-477

Lahiri. T., Roy, S., Basu, C., Ganguly, S., Ray, M. R. and Lahiri, P. (2000) Air pollution in Calcutta elicits adverse pulmonary reaction in children, *Indian Journal of Medical Research*, 112,:21-26.

Lakshmanan , T.R.(2004) Air quality in Asian megacities: issues and policies, Working paper series, Centre for transportation, Boston University, Working Paper # CTS2004D http://www.bu.edu/transportation/WPSeries.html

Lee, J.-T., Kim, H., Hoh, Y.-C.,Kwon, H.-J., Schwartz, J and Christiani, D. C. (2000) Air pollution and daily mortality in seven major citites of Korea, 1991-1997, *Environmental Research,* A84:247-254

Leontief. W (1951) *The Structure of American Economy, 1919-39*, Second Edition, Oxford University Press, New York.

Liaskas, K., G. Mavrotas, M. Mandaraka and D. Diakoulaki, (2000) Decomposition of industrial CO_2 emissions: the case of European Union, *Energy Economics*, 22: 383– 394.

Lin X. and Polenske, K. R. (1995) Input-Output anatomy of China's energy use: changes in the 1980, *Economic Systems Research*, 7(1): 67-83.

Lin, G., (1998) Energy development and environmental constraints in China, *Energy Policy*, 26 (2):119– 128.

Lin, S. and Chang, T.C., (1996) Decomposition of CO_2, SO_2 and NOx emissions from energy use of major economic sectors in Taiwan, *Energy Journal* 17 (1):1 –17.

Lioy, K (1990) Assessing total human exposure to contaminants, *Environmental Science and Technology* 24(7): 938–945

Lippmann M (2003) 'Air pollution and health – studies in the Americas and Europe', In Air Pollution and Health in Rapidly Developing Countries edited by Frank Murray and Gordon McGranahan, Earthscan, London

Loomis, D, Castillejos, M, Gold, D R et al (1999) Air pollution and infant mortality in Mexico City, *Epidemiology*, 10 :118-123.

Lvovsky K (1998) Economic costs of air pollution with special reference to India, South Asia Environment Unit, World Bank, Prepared for the National Conference on Health and Environment, Delhi, India, July 7-9, 1998

Maenpaa, I., (1998) The economy, energy and air emissions, Working Paper No 2 / 1998/ b2, November Statistics Finland EURO Stat.

Mage, D., Ozolins, G., Peterson, P., Webster, A., Orthofer, R., Vandeveerd, V., Gwynne, M.(1996) Urban air pollution in megacities of the world, *Atmospheric Environment* 30:681-686.

Martins M.C.H., Fatigati F.L.,Vespoli T.C, Martins L.C, Pereira L.A.A., Martins M.A., Saldiva P.H.N and Braga A.L.F.(2004) Influence of socioeconomic conditions on air pollution: Adverse health effcts in elderly people: An analysis of six regions SaoPaulo, Brazil, *Journal of Epidemiology and Community Health*, 58:41-46

Mashelkar, R. A., Biswas, D. K., Mathur, O. P., Natarajan, R., Nyati, K. P., Singh, D. V. and Singhal, S. (2002) Report of the expert committee on auto fuel policy, Government of India, New Delhi, http://petroleum.nic.inl.

McGranahan, G. and Murray, F. edited (1999) Health and air pollution in rapidly developing countries, Stockholm Environment Institute, Sweden.

Mendelsohn, R., Orcutt, G. (1979) An empirical analysis of air pollution dose-response curves, *Journal of Environmental Economics and Management* 6(2):85-106

Miller, R. and P. Blair(1985) *Input-Output analysis: foundations and extensions*. Englewood Cliffs, New Jersey: Prentice-Hall.

MoEF(2002) The Air Prevention and Control of Pollution Act, 1981 Ministry Of Environment and Forest, Government of India

Mohanraj ,R and P. A. Azeez (2004) Health effects of airborne particulate matter and the Indian scenario, *Current Science*, 87(6), 25 September

Molina Mario J. and Molina Luisa T. (2004) Megacities and atmospheric pollution, *J. Air and Waste Manage. Assoc.* 54:644–680

Moran, D. et al., (2000) Economics and environmental health: A case study valuing morbidity impacts of urban air pollution in Malaysia, WHO publication

Mukhopadhyay, K., Chakraborty, D., (1999) India's energy consumption changes during 1973–1974 to 1991–1992, *Economic Systems Research* 11 (4): 423– 437.

Mukhopadhyay, K (2002) A Structural decomposition analysis of air pollution from fossil fuel combustion in India, *International Journal of Environment and Pollution* 18(5): 486-497

Mukhopadhyay, K (2002a) Energy consumption changes and CO_2 emissions in India, New Delhi: Allied Publishers.

Mukhopadhyay, K and Forssell O. (2005). An empirical investigation of air pollution from fossil fuel combustion and its impact on health in India during 1973-74 to 1996-97, *Ecological Economics*, 55(2):235-250

Mullahy, J., and Portney, P. (1990) Air pollution, cigarette smoking, and the production of respiratory health, *Journal of Health Economics* 9(2): 193-205.

Muneer T, Muhammad A, and S. Munawwar (2005) Sustainable production of solar electricity with particular reference to the Indian economy, *Renewable and Sustainable Energy Reviews*, 9 (2005) :444–473

Murray,F, Mcgranahan G., Kuylenstierna J.C.I.(2001) Assessing health effects of air pollution in developing countries, *Water air and soil pollution*,130(1-4): 1799-1804 August

Murray,F and McGranahan G, (2003) "Air pollution and health in rapidly developing countries-the Context" In *Air pollution and Health in Developing countries* edited by McGranahan Gordon and Murray Frank, Earthscan Publications Limited, London U.K.

Murty. M. N., Gulati. S. and Mukherjee. A (2002) Hedonic property prices and valuation of benefits from reducing urban air pollution in India, Institute of Economic Growth *Working paper series*

Narain, S.(2002), Deceptive asthma. *Health Environ.*, 1, 1–16.

NSSO (1984).Consumer Expenditure (1983), Schedule no. 1.0, 38th round, CD# 2100and2103, National Sample Survey Organisation, Ministry of Statistics and Programme Implementation, New Delhi

NSSO (1991).Consumer Expenditure (1989-90), Schedule no. 1.0, 45th round, CD# 1868, National Sample Survey Organisation, Ministry of Statistics and Programme Implementation, New Delhi

NSSO (1995).Consumer Expenditure (1993-94), Schedule no. 1.0, 50th round, CD# 2243,2244 and2245, National Sample Survey Organisation , Ministry of Statistics and Programme Implementation, New Delhi

NSSO (2001).Consumer Expenditure (1999-2000), Schedule no. 1.0, 55th round, CD# 2942, National Sample Survey Organisation, Ministry of Statistics and Programme Implementation, New Delhi

NSSO (2004) Household Consumer Expenditure in India, Schedule no. 1.0, 60th Round, National Sample Survey Organisation (January–June 2004) Report No. 505, Chapter Three

NEERI, (2004) Long term study of air quality trends in major urban centres, Mumbai, zonal centre, national environmental engineering research institute of Delhi. *Atmospheric Environment*, 1992, 26B :291-294.

Ostro, B, Chestnut, L, Vichit- Yadakan, and Laixuthai, A (1999) 'The impact of fine particulate matter on daily mortality in Bangkok, Thailand' in J Chow et al., edited $PM_{[2.5]}$: A Fine Particulate Standard, vol II, Long Beach, CA, :939-949.

Ostro, B., et al., (1995) Air pollution and mortality: results from a study of Santiago, Chile, *Journal of Exposure Analysis and Environmental Epidemiology*, 6: 97-114.

Pachauri S(2004) An analysis of cross-sectional variations in total household energy requirements in India using micro survey data, *Energy Policy* 32 :1723–1735

Pachauri, S., Spreng, D (2002) Direct and indirect energy requirements of households in India, *Energy Policy* 30 (6): 511–523.

Pandey J. S., R. Kumar, S. Devotta(2005) Health risks of NO2, SPM and SO2 in Delhi (India) *Atmospheric Environment* 39 : 6868–6874

Parikh, J.(2001) Poverty-environment Nexus, *International Journal of Global Environmental Issues*, 1(1).

Parikh, J. K. and Hadker, N (2003) Economic impacts of urban air Pollution: valuation for Mumbai, India, *International Journal of Environment and Pollution*, 19:498-515.

Paramesh, H. and Cherian, E., (2001) Bangalore besieged. *Down to Earth*, : 48–49.

PCD(2000) Air emission source database update 2000, Bangkok, Thailand, Pollution Control Department, www.pcd.go.th

Pearce D, (1996) Economic valuation and health damage from air pollution in the developing world, *Energy Policy*, 24(7): 627-630

Pereira, L A, Loomis, D, Conceicao, G M, Braga, A L F, Arcas, R M, Kishi, H S, Singer, J M, Bohm, G M and Saldiva, P H (1998) Association between air pollution and intrauterine mortality in Sao Paulo, Brazil, *Environmental Health Perspectives*, 106,:325-329

Pino, P, Romieu, I, Villegas, R, Walter, T and Oyarzun, M (1998) Wheezing bronchitis in infants: influence of airborne fine particulate matter and environmental tabacco smoke exposure, *American Journal of Respiratory Critical Care and Medicine*, A877.

Pope C.A. III, M.J. Thun, N.M. Nmboodiri, D.W. Dockery, J.S. Evans, F.E. Speizer and C.W. Heath, (1995), Particulate air pollution as a predictor of mortality in a prospective study of US adults, *American Journal of Respiratory Critical Care Medicine* 151 : 669–674

Pope, C.A. II. , D.W. Dockery, J.D. Spengler and M. E. Razienne (1991) Respiratory health and PM10 pollution : A daily time series analysis, *American Review Respiratory Disease.* 144: 688-674.

Pope, C.A.II., J.Schwartz and M. Ransom(1992), .Daily mortality and PM10 pollution in Utha Valley, *Archives in Environmental Health*, 47 (3) : 211-217.

Radian International LCC (1998) Particulate Matter abatement strategy for the Bangkok metropolitan area, Bangkok, Thailand, Pollution Control Department, Ministry of Science, Technology and Environment

Ravindra K; Eric Wauters, Sushil K. Tyagi, Suman Mor and Ren'E Van Grieken.(2006) Assessment of air quality after the implementation of compressed natural gas (cng) as fuel in public transport in Delhi, India, *Environmental Monitoring and Assessment* 115: 405–417

Reddy Y V (2007) Glimpses of the Indian economy and its financial sector, Speech by Dr Y V Reddy, Governor of the Reserve Bank of India, at the Central Bank of the Russian Federation, Moscow, 2 July 2007

Resosudarmo Budy P. and Lucentezza Napitupulu (2004) Health and economic impact of air pollution in Jakarta, *The Economic Record*, 80, Special Issue, September :S65–S75

Ritz, B, Yu, F, Chapa, G and Fruin, S (2000) Effect of air pollution on preterm birth among children in southern California between 1989 and 1993, *Epidmllology*, 11(5), :502-511.

Romero-Placeres M, Mas-Bermejo P, Lacasana-Navarro M, Tellez Rojo-Solis MM, Aguilar-Valdes J, Romieu I.(2004) Air pollution, bronchial asthma, and acute respiratory infections in minors, Habana City [in Spanish]. Salud Publica Mex;46:222-233.

Romieu I(1999) 'Epidemilogical studies of health effects arising from motor vehicle air pollution' in Urben Traffic Pollution, Ecotox/WHO/E& F N Spon, edited by D. Schwella and O. Zali, London and New York :9-49

Romieu I and M H Avila (2003) 'Air pollution and health in developing countries – A review of epidemiological evidence'; In *Air Pollution and Health in Rapidly Developing Countries* edited by Frank Murray and Gordon McGranahan, Earthscan Publications, London

Romieu, I, Meneses, F. Ruiz, S, Huerta, J, Sienra, J J, White, M, Etzel, R and Hernandez Avila, M (1997) Effects of intermittent ozone exposure on peak expiratory flow and respiratory symptoms among asthmatic children in Mexico City, *Archives of Environmental Health*, 52(5):368-376

Romieu, I and Hernandez, M., (1999) 'Air pollution and health in developing countries: review of epidemiological evidence', In *Health and Air Pollution in Rapidly Developing Countries* edited by Mc Granahan, G., Murray, F., Stockholm Environment Institute, Sweden, 43– 66.

Rose A, and Chen CY (1991) Sources of change in energy use in the US economy 1972–1982, *Resource Energy* 13: 1–21

Roy, S., Ray, M. R., Basu, C., Lahiri, P. and Lahiri, T. (2001) Abundance of siderophages in sputum. indicator of an adverse lung reaction to air pollution, *Actaytologica*, 45:958-964.

Saksena. S and K R. Smith (2003) 'Indoor air pollution', In *Air Pollution and Health in Rapidly Developing Countries* edited by Frank Murray and Gordon McGranahan, Earthscan, London

Saldiva, P H N, Lichtenfels, A J R C, and Paiva, P S (1994) Association between air pollution and mortality due to respiratory diseases in children in Sao Paulo, Brazil: a preliminary report, *Environmental Research* 65:218-25

Samet JM, F Dominici, FC Curriero, I Coursac and SL Zeger (2000) Fine particulate air pollution and mortality in 20 U.S. cities, 1987–1994, *N Engl. J Medicine*, 343 (24):1742-1749

Schwartz J. (1991), .Particulate air pollution and daily mortality in Detroit, *Environmental Research* , 62, 7-13.

Schwartz, J., and D.W. Dockery. (1992a) Increased mortality in Philadelphia associated with daily air pollution concentrations, *American Review of Respiratory Disease* 145:600-604.

Schwartz, J., and D.W. Dockery. (1992b) Particulate air pollution and daily mortality in Steubenville, Ohio, *American Journal of Epidemiology* 135(1):12-19.

Schwartz, J. (1993), .Particulate air pollution and chronic respiratory disease, *Environmental Research,* 56: 204-213.

Schwartz, J. and Morris, R. (1995) Air pollution and hospital admissions for cardiovascular disease in Detroit, Michigan, *American. Journal.of Epidemiology* 142:23-35.

Schwartz, J. et al., (1996) Is daily mortality associated specifically with fine particles? *Journal of the air and waste management association,* 46(10): 927–939

Schwela ,D(2000) Air pollution and health in urban areas, *Reviews on Environmental Health* 15 (1–2): 13–42

Schwela, D., Haq, G., Huizenga, C., Han, W., Fabian, H., and Ajero, M. (2006) Urban air pollution in Asian cities status, challenges and management, Earthscan Publishers, London:224p

Senayake, M. P., Samarakkody, R. P., Jasinghe, S. R., Hettiarachchi. A. P., Sumanasena, S. P. an Kudalugodaarachchi, J. (1999) Association between ambient air pollution and acute childhood wheezy episodes in Colombo', Forestry and Environment Symposium, Department of Forestry and Environmental Science, University of Sri Jayewardenpura, Sri Lanka, http://ybiol.tripod.com/forest/99sympo/9924sena.htm

Sharma, A.K. et al., (1999) Descriptive epidemiology of acute respiratory infections among under five children in an urban slum area, *Indian Journal of Public Health,* 43(4).

Shirazi, M.A. and Harding, A.K. (2001). Ambient air quality levels in Tehran, Iran, from 1988 to 1993, *International Journal of Environment and Pollution,* 15 (5):517-528

SIDA (2006) Urban issue paper, supplement to SIDA's policy for urban development, fighting poverty in an urban world, www.sida.se/urban

Sikdar, P.K. and Mondal, S. (1999). Air pollution management programme of Calcutta, *Journal of Environmental Studies and Policy,* 2 (2): 71-86.

Singh, A.; Sarin, S.M.; Shanmugam, P.; Sharma, N.; Attri, A.K.; Jain V.K. (1997) Ozone distribution in the urban environment of Delhi during winter months, *Atmos. Environ.* 31(20):3421-3427.

Sinha, S. and Bandhopadhyay, T.K. (1998). Review of trace elements in air environment and its health impact in some Indian cities, *Journal of IPHE,* India, 1: 35-48.

Smith KR, and Liu Y. (1996)Indoor air pollution in developing countries. In: Samet JM, editor. *Epidemiology of Lung Cancer*. New York: Marcel Dekker; : 151-79.

Smith K.R. et al.,(1999) How much global ill health is attributable to environmental factors?, Epidemiology **10** (1999) (5):573–584

Srivastava, A., and Kumar, R. (2002) Economic valuation of health impacts of air pollution in Mumbai, *Journal of Environmental Monitoring and Assessment,* 75(2): 135-143.

Stern R.E. (2003) Hong Kong Haze: air pollution as a social class issue, *Asian survey* 43(5):780

Tamura, K., Jinsart, W., Yano, E., Karita, K and Boudoung, D. (2003) Particulate air pollurion and chronic respiratory symptoms among traffic policemen in Bangkok, *Archives of Environmental Health*, 58:210-207

TEDDY (various issues) TERI Energy Data Directory & Yearbook, New Delhi:TERI

Tellez-Rojo, M. M. Romieu, I, Pena, M P, Ruiz-Velasco, S, Meneses-Gonzales, F and Hernandez-Avila, .M (1997) Efecto de la contaminacion ambiental sobre las consultas por infecciones respiratorias en ninos de la Ciudad de Mexico, *Saland Publica de Mexico*, 39(6):513

Tellez-Rojo, MM, Romieu, I, Ruiz-Velasco, S, Lezana, M A and Hernandez-Avila, M.M.(2000) Daily respiratory mortality and PM_{10} pollution in Mexico City: importance of considering place of death, *European Respiratory Journal*, 16:391-396.

TERI (1997) Air pollution and health in India, TEDDY, TERI Publication New Delhi.

TERI (2001),The Energy and Resources Institute. Review of past and on-going work on urban air quality in India; TERI Report No. 2001EE41; TERI: New Delhi

TERI (2001a)The Energy and Resources Institute. State of Environment Report for The Case of Delhi. Report No. 2000EE65, TERI, New Delhi.

Thang,Q.Dam (2004) A vision for cleaner emissions from motorcycles in Vietnam, Paper presented at the cleaner vehicles and fuels in Vietnam Workshop, 13-14 May, Hanoi, Vietnam

Torres, E Ronald D. Subida, R.D.,Gapas, JL.,Sarol J.N.Villarin J.T.,Vinluan, R.J.N.,Ramos B.M. and Quirit LL(2004) Public health monitoring : a study under the Metro Manila air quality improvement sector development program (MMAQISDP), Manila ADB.

Transport Department (2004) Government of NCT of Delhi, New Delhi, http://transport.delhigovt. nic.in/.

UNEP(1991) Urban Air Pollution, United Nation Environment Programme, Nairobi

UNEP /WHO (1992) Urban Air Pollution In Megacities of the World, United Nations Environment Programme/ World Health Orgainsation, Oxford, Blackwell.

UNEP(2001) India – State of Environment , United Nations Environment Programme, ESCAP Thailand, : 97–112.

Varshney, C.K; Aggarwal, M (1992) Ozone pollution in the urban environment of Delhi. *Atmospheric. Environment*,26B:291-294.

Viet Nam Register (2002) Integrated action plan to reduce vehicle emissions in Viet Nam. Asian Development Bank, Manila. www.adb.org/Vehicle-Emissions/actionviet.asp

Vietnama Panorama News Online(2005) Health environment. Hanoi facing severe air pollution, Vietnam panorama news online, 30th November

Von Schirnding, Y E R, Yach, D, Blignaut, R and Mathews, C (1991) Environmental determinants of acute respiratory symptoms and diarrhea in young coloured children living in urban and peri-urban areas of south Africa, *South African Medical Journal*, 79(8) :457-461

Walsh, M. P,(2003) Vehicle emissions trends and forecasts the lessons of the past 50 years, Blue Sky in the 21st Century Conference, Seoul. Korea

Walter, P., and Fellner, W. (1978) Exploratory techniques for the determination of potential dose-response relationships between human health and air pollution, *Journal of Environmental Economics and Management* 5(4): 376-89.

Wang, X., Ding, H., Ryan, L. and Xu, X. (1997) Association between air pollution and low birth weight: A community based study, *Environmental Health Prespectives*, 105(5) :514-520

WBDoE(2003) Air quality and ambient air quality and sources of pollution, Department of Environment, Government of West Bengal

WBPCB (2001) Green governance newsletter of West Bengal Pollution Control Board, II (III) December

WBPCB (2003) A Quinquenniel Report April,1998-March 2003, Kolkata India, West Bengal Pollution Control Board

WBPCB (2007) Pollution Report in West Bengal, West Bengal Pollution Control Board

WHO (2000a) Guidelines for Air Quality, WHO/SDE/OEH/00.02, Geneva, World Health Organisation

WHO (2000b) Air Quality Guidelines for Europe, (World Health Organization), Regional Office for Europe, Copenhagen.

WHO (2002) World Health Organisation Report, World Bank publication

WHO (2003) Health aspects of air pollution with Particulate Matter, Ozone and Nitrogen dioxide, Report of a WHO working group Bonn, 13-15 January, www.who.dk/document/e79097.pdf

WHO (2006) Use of The Air Quality Guidelines in Protecting Public Health: A Global Update, Fact Sheet No313, October, World Health Organisation, Geneva

WHO(2006a) Health impact assessment methods and strategies, Health impact of PM_{10} and ozone in 13 Italian cities, http://www.euro.who.int/healthimpact /MainActs/20020619_2

Wier, M., (1998), Sources of changes in emissions from energy: a structural decomposition analysis, *Economic Systems Research* 10 (2):99– 112.

Wirahadikusumah,K (2002) Jakarta air quality management: trends and policies, Presentation at the Regional Workshop on better air quality in Asian and Pacific Rim cities, Hong Kong 16-18 December

Woodruff. T, Grillo, J and Schoendorf, K G (1997) The relationship between selected causes of postneonatal infant mortality and particulate air pollution in the United States, *Environmental Health Perspective*, 10(6) : 608-612.

World Bank (1992) World Development Report, Development and the Environment, 1, World Bank publication, Washington D.C.

World Bank (1997) Urban air quality management strategy in Asia guidebook, URBAIR, Edited by Jitendra J. Shah Tanvi Nagpal and Carter J. Brandon, Washington, D.C.

World Bank (1997a) Urban air quality management strategy in Asia: Greater Mumbai Report, *World Bank technical paper No. 381*, Washington DC, World Bank.

World Bank(2000) Urban air quality management the transport- environment-energy nexus, prepared by Masami Kojima and Magda Lovei (Oil, Gas, and Chemicals Department and Environment Department of the World Bank), April

World Bank(2001) Project appraisal document for the metro Manila urban transport integration project, www.gefweb.org/

World Bank (2002) Thailand environment monitors 2002 for air quality, Washington DC World Bank, www.worldbank.org/

World Bank (2004) Toward cleaner urban air in South Asia: tackling transport pollution, understanding sources, Washington Dc. Joint UNDP/ World Bank Energy Sector Management Assistance Programme (ESMAP)

World Development Indicators (2000) World Bank publication, Oxford University Press, USA.

World Development Indicators (2007), WDI online databases

World Development Report (2005) World Bank publication, Washington DC World Bank, www.worldbank.org/

WRI (1998-99) China's health and environment, air pollution and health effects, in *World Resources 1998-99: Environmental change and human health*, World Resource Institute, A joint publication by the World Resources Institute, the United Nations Environment Programme, the United Nations Development Programme, and The World Bank. 1998. www.thedailystar.netl2oo3/07/20/d30n0250278.htm

Xie, Jian, C. J. Brandon and J. J. Shah.(1998) Fighting urban transport air pollution for local and global good : the case of two-stroke engine three -wheelers in Dhaka.. Presented at the Integrated Approach to Vehicular Air pollution in Dhaka Consultative Meeting, Bangladesh, April 26-27

Xu. X., Gao, J., Dockery, D. W. and Chen, Y. (1994) Ar pollution and daily mortality in residential areas of Beijing, China, *Archives of Environmental Health*, 49(4) : 216-222

Xu, X., Ding, H. and Wang, X. (1995a) Acute effects of TSP and SO2 on preterm delivery: A community based cohort study, *Archives of Environmental Health*, 50(6) :407-415

Xu, X., Christiani, D. C., Li, B. and Huang, H. (1995b) Association of air Pollution with hospital outpatient visits in Beijing, *Archives of Environmental Health*, 50,(3): 214-220

Yang, C.Y. Chang C.C., Chuang, H.Y.,Tsai, S.S., Wu, T.N. and Ho, C.K. (2004) Relationship between air pollution and daily mortality in subtropical city Taipei, Taiwan, *Environment International*,30(4): 519-523

Yu, T (2002) Air quality monitoring in Beijing, Proceedings of the better air quality Workshop 2002, Hong Kong, 16-18 December

Zhang X.Z. and H. Folmer, (1998) Economic modeling approaches to cost estimates for the control of CO_2 Emissions: a case study of Germany and U.K., *Energy Economics*, 20(1): 101-120.

Zuidema,T. and Nentjes (1997) Health damage of air pollution: An estimate of a dose response relationship for the Netherlands. *Environmental and Resource Economics,* 9 (3):291-308.

INDEX

A

B

C

D

Q

R

S

T

U

V

W

Z